Action Learning and Action Research Journal

Vol 28 No 1 November 2022

Action Learning, Action Research Association Ltd (and its predecessors) has published the ALAR Journal since 1996.

Managing Editor: Dr Yedida Bessemer

Issue Editor: Akihiro Ogawa

Global Strategic Publications Editorial Board:
Dr Azril Bacal, University of Uppsala, Sweden
Dr Susan Goff, Murray-Darling Basin Authority, Australia
A/Prof Erik Lindhult, Mälardalen University, Sweden
Riripeti Reedy, Ngati Porou, Director, Maitai Group Ltd, New Zealand
Prof Shankar Sankaran, University of Technology Sydney, Australia

Editorial inquiries:
The Editor, *ALAR Journal*
Action Learning, Action Research Association Ltd
PO Box 162, Greenslopes, Qld 4120 Australia

editor@alarassociation.org

ISSN 1326-964X (Print)   ISSN 2206-611X (Online)

The Action Learning and Action Research Journal is listed in:

- Australian Research Council – *Excellence in Research for Australia (ERA) 2018 Journal List*
- Australian Business Deans Council - *2019 ABDC Journal Quality List*
- Norwegian Directorate for Higher Education and Skills - *Norwegian Register for Scientific Journals, Series and Publishers*

Editorial Advisory Board

Dr Yedida Bessemer	USA
Dr Gina Blackberry	Australia
Dr Deeanna Burleson	USA
Dr Daniela Cialfi	Italy
Dr Ross Colliver	Australia
Mr Andrew Cook	Australia
Dr Carol Cutler White	USA
Capt Michael Dent	Malaysia
Dr Bob Dick	Australia
Dr Susan Goff	Australia
Assoc. Prof. Marina Harvey	Australia
Dr Geof Hill	Australia
Ms Jane Holloway	Australia
Dr Marie Huxtable	UK
Dr Angela James	South Africa
Dr Brian Jennings	Ghana
Dr Diane Kalendra	Australia
Prof. Vasudha Kamat	India
Prof. Nene Ernest Khalema	South Africa
Dr Elyssebeth Leigh	Australia
Dr Tome Mapotse	South Africa
Dr John Molineux	Australia
Ms Margaret O'Connell	Australia
Dr Terry Parminter	New Zealand
Dr Paul Pettigrew	UK
Dr Eileen Piggot-Irvine	New Zealand
Dr Franz Rauch	Austria
Dr Michelle Redman-MacLaren	Australia
Ms Riripeti Reedy	New Zealand
Prof Wendy Rowe	Canada
Dr Akihiro Saito	Japan
Prof Shankar Sankaran	Australia
Assoc. Prof. Sandro Serpa	Portugal

Dr Steve Smith	Australia
Prof Emmanuel Tetteh	USA
Dr Vicki Vaartjes	Australia
Prof Jack Whitehead	UK
Assoc Prof Hilary Whitehouse	Australia
Dr Janette Young	Australia

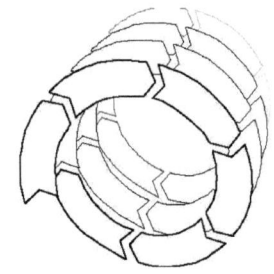

ALAR Journal

Volume 28 No 1
November 2022

ISSN 1326-964X (Print)
ISSN 2206-611X (Online)

CONTENTS

Editorial	7
Yedida Bessemer	
Foreword for special issue	13
Davydd J. Greenwood	
Introduction: Learning, practicing, connecting, and passing along	16
Akihiro Ogawa	
A 'veteran reflects' – A personal biographic response to Akihiro Ogawa's Introduction	39
Yoland Wadsworth	
Research reflexivity: A journey of unlearning and co-learning	51
Ai Ming Chow	

'She might find out the truth:' Action researching with young theatre audiences — 72

Abbie Trott

An HRM student's search for relevancy — 99

Edward Hyatt

Researching resilience in action: Matters of action research as 'matters of care' — 126

Akina Mikami

Conclusion: Walking beside co-researchers and finding off ramps in PhD dissertation journeys — 153

Ai Ming Chow, Ed Hyatt, Akina Mikami, Asha Ross, Abbie Trott

Membership information and article submissions — 182

© 2022. Action Learning, Action Research Association Ltd and the author(s) jointly hold the copyright of *ALAR Journal* articles.

Editorial

Greetings,

Since the phrase action research (AR) was first coined by Kurt Lewin in the 1940s (Lewin, 1946), it has evolved somewhat. This ALARj special issue truly is special, including as it does several stories of AR learning journeys across three academic generations. The multi-generational authors invite readers to learn from their own research experiences and reflections and to draw from them a greater understanding of the significance of action research.

In the foreword, Davydd Greenwood shares a brief account of one of his students, Akihiro Ogawa, and his pursuit of action research at Cornell University some two and a half decades ago. Greenwood supported Ogawa in his path and praised Ogawa's dedication and passion for action research, and even more so, for his efforts to pass his learning to his own students.

The first paper is that of Akihiro Ogawa, in which he describes and explains in greater detail his AR learning experience from his graduate student days until the present, where he is currently teaching AR to the next generation of students. Ogawa's account, which is written in chronological order, is divided into four sections. He began with his first AR encounter at Cornell University and how he was fascinated by AR as a graduate student, something which led him to employ AR methodology in his doctoral study. He then explains his work and AR practice at a non-profit organization (NPO) in Japan, where he also volunteered and conducted his doctoral fieldwork; this work served as a fertile ground for AR practice and learning. Later, he connected with AR scholars and practitioners. Finally, he developed an AR seminar for students, through which he instilled in his students the knowledge and skills to conduct an AR. Moreover, the seminar participants, beyond understanding the importance of action research and its relevance and value to people and cultures, then had to put what they had learned into practice.

Next, Yoland Wadsworth offers a veteran's response to Ogawa's AR odyssey. Unlike Ogawa, who "*discovered*" AR through books at Cornell University Library, Wadsworth, an experienced AR practitioner (over four decades), shared her AR journey from a time when there were no available AR books or AR "glossary"; in other words, she was among the AR pioneers in the academia. Wadsworth also divides her article into four sections, each focusing on a period from the early 1970s until today; each description and explanation corresponds with the needs and the environment of its time.

Wadsworth started with the need to connect with other scholars who experienced similar motivations to move from traditional positivism. Next, she and her colleagues practiced AR through trial and error on different research projects. Then came the stage to pass their AR knowledge and skills to others, and they did so by connecting globally with prominent worldwide scholars. Wadsworth pointed out in the last section the need for deeper learning and inquiry while putting humanity at the center.

It is important to note that even though Ogawa and Wadsworth's AR journeys were distinctive to their generations, they both insisted on passing on the practice of AR. Wadsworth and Greenwood passed their knowledge to the next generation, Ogawa, who received the "AR baton," as Wadsworth described it, and took it further by creating an AR seminar to help Ph.D. The following five papers demonstrate this valuable work of the younger academic generation. The first four papers are individual doctoral AR project reflections; Chow and Trott's studies relate to the art field, while Hyatt examined human resource management practice, and Mikami researched the resilience and healing of a small community stricken by a disaster. The last paper is a cooperative reflection of five postgraduate students (the fifth student decided to withdraw her individual reflection paper from the journal), reaching a culmination of their work.

Ai Ming Chow's paper is the first doctoral reflection, and it focuses on the indigenous art markets in Australia by conducting participatory action research (PAR) within this community. She indicated that she is not Australian or of indigenous descent and had difficulties connecting with the people in this community before her study. After reading and learning about the indigenous history and culture, she gained trust and formed relationships.

Chow examined the diverse stakeholder groups in the indigenous communities and asserted that to correct the injustice caused by colonization; there is a need for a process of *un-learning* that includes relinquishing power. Then, through her research work with the different groups and formal and informal fieldwork, Chow could outline a *co-learning* process with them. Moreover, Chow encourages us to acknowledge others' frames of reference, keep an open mind, and have humility when learning and researching other cultures, such as the indigenous communities in Australia and, more specifically, their art markets' groups. Doing so will help empower the indigenous people to voice their opinions and share their rich culture and knowledge with the world.

Like Chow, Abbie Trott's paper examined the art field, specifically theatre and performance. She inquired about how young people, primarily teenagers, are involved with the theatre when digital media is global and crosses boundaries. Using AR, Trott ensured the active participation and empowerment of the young people in the study. She focused on four performance case studies where the young people engaged with them differently, and the path changed according to the audience. Trott pointed out the importance of understanding the audience and allowing the participants to direct the way; she calls on future theatremakers to plan and act accordingly, considering the audience's experiences and the participants.

Edward Hyatt's is the third doctoral reflection paper, and it examines the human resource management (HRM) field, particularly the job interview in the employee selection process. Despite his initial hesitation to conduct a doctoral AR, Hyatt eventually discovered that AR principles align with him as a researcher and person. He claimed that there is a gap between business schools and practitioners since organizations refuse to implement research findings that they find irrelevant. Hyatt's goal was to bridge this divide by employing AR and creating a better job interview practice to improve the prediction of applicants' performance.

Hyatt ultimately had to abandon his experimental design in the face of his supervisors' rejection of it, but there was a silver lining for him as he changed his research direction. He found confirmation in the literature of his initial analysis of the flaw in the traditional HRM system and supported his critique. In addition, his research philosophy and design approach correlated with AR, reinforcing that interaction with the participants. Hyatt's message to researchers was to embody humility and a problem-solving mindset and to seek more participatory efforts to link HRM scholarship and business.

Akina Mikami's paper is the fourth doctoral reflection. Mikami studied resilience care and recovery practices in the wake of a disaster, specifically the Fukushima group of children affected by the Tōhoku earthquake, and the tsunami that triggered the Tokyo Electric Power Company's (TEPCO) nuclear power plant accident on March 11, 2011. Mikami's grassroots research study attempted to allow children who live in disaster areas affected by radiation to clear their minds and refresh their bodies in different locations during school holidays and weekends.

Mikami conducted a five-year research study and engaged in a collaborative inquiry that examined what it means to *do* resilience in civil society in Cairns, Australia. Using AR methodology, she gained insights into how civil society forms alternate thoughts and views of resilience and healing. Mikami pointed out the importance of letting the Fukushima children voice their opinions, impressions, and challenges about their time outdoors and experiences and listening to their real-life stories to create a better future for them.

All the authors shared their backgrounds in order to provide a frame of reference, the challenges they encountered, and the lessons they learned. Moreover, they shared their reflexive accounts of their research journeys and allowed us, the readers, to join them in this enriching world of action research. These personal stories can motivate us to take action, improve, and share it with the world.

The last paper in this special issue is titled "Conclusion: Walking beside co-researchers and finding off ramps in Ph.D. dissertations journeys." Ai Ming Chow, Ed Hyatt, Akina Mikami, Asha Ross (who decided to withdraw her doctoral reflection paper from this issue), and Abbie Trott collaborated and co-wrote this article. In it, they discuss employing AR methodology in doctoral research to help future students interested in applying AR. After providing background of their studies and collaborative writing process, the five authors structured their reflections by addressing the following topics: the distinction between AR and traditional research, the way to manage research ethics, utilizing AR in a Ph.D. timeline, and expressing and understanding AR philosophical principles in doctoral dissertations. The authors also add a section addressing how their philosophical approach to AR manifested in their doctoral projects. Lastly, they concluded with their hope that their AR learning journey reflections would contribute to the next generation of students just as the previous one paved the way for them as if they were passing a family heirloom.

Just like the "first academic generation" authors, the "second" and "third" ones hope to make an impact and initiate a change to improve the lives of the people who participated in their research and contribute to their field of study. It seems that the seeds planted three decades ago at Cornell University have come to fruition in other lands far across the ocean are expressed in this journal issue.

To Lifelong Learning,

Dr. Yedida Bessemer

References:

Lewin, K. (1946). Action research and minority problems. *Journal of Social Issues, 2*(4), 34-46. doi: 10.1111/j.1540-4560.1946.tb02295.x

Foreword for special issue
Davydd J. Greenwood

Received May 2022   Reviewed August 2022   Published November 2022

Over 44 years at Cornell University, I had the good fortune to teach and learn from 100s of wonderful students. Of these, Akihiro Ogawa is one of the most memorable. He found me, not the other way around. When he did, he told me what he intended to do. Ogawa had decided that Action Research (AR) was what he wanted to learn, that he wanted to switch graduate fields to Anthropology to do that and to do it all in record time. I remember thinking that it would be challenging for him to succeed at this, but I made the wise decision to support his efforts, along with our late colleague, Ted Bestor. The rest is, as they say, "history." Not only did he achieve his goal, but he produced a fine dissertation and has built an international career as an action researcher in an academic world that is not welcoming of such efforts. The moral of this story is to let smart people find their own way and figure out how to support their efforts. We don't need more academic clones. We need more people like him.

The reader will have noticed how uniquely personal and linked to individual strengths, knowledge, and experiences Ogawa's AR research and teaching practice, as is Yoland Wadsworth's, mine, and this new generation of students. I am heartened by the way Ogawa has taken on the task of sewing together transgenerational networks of AR to link the older generations to the coming generations, all without demanding orthodoxies. He is creating mutually supportive networks of co-learners.

What stands out in the texts that follow is the way each of us has pursued an idiosyncratic path into AR. We found our way to this practice out of a combination of epistemological, political, and ethical dissatisfaction with social science business as usual. This also means it is not a "school." That would defeat the very essence of what makes each individual uniquely qualified to practice a particular kind of AR in a particular context and with a specific set of social aims. AR liberates the unique talents and experiences of each practitioner, something fully evident in the cases and the conclusions the students have written to this special issue. No further proof of the value and effectiveness of this concept is needed beyond the thoughtful and thorough conclusion the former students have fashioned together. It is as good a synthesis of AR going forward as I have ever seen.

Being old and dispirited by the ongoing planetary ills of massive global inequality, racism, sexism, ecological destruction, and now the threat of world war yet again, I read this special issue as a testimony to the viability of a better path toward human and planetary flourishing. We clearly know that AR works produce meaningful social scientific knowledge and practical and collaborative relationships between researchers and local stakeholders capable of improving the quality of human life.

We also know it is not a dominant practice anywhere. Why not? Since the founding of the social sciences at the end of the 18th century, they were premised on learning the "laws" governing human societies for the purpose of improving those societies. How did we come to accept the idea of social science as an objective, non-applied set of unrelated "disciplinary" practices? It is not an accident.

Disciplining academic social researchers to stay away from the real world has taken many forms, from active censorship, academic expulsions, and, most recently, the application of the "New Public Management" to social research. Armies of "accountants" with spreadsheets count and rank social research products as if they were car parts or army boot inventories. This anti-social strategy gives the "rankers" the power to decide what counts. It has been deployed to ensure that most academic social scientists will not trouble financial and political elites, polluters, racist groups, and other predators with their research projects and results.

Beyond this, the next challenge is already with us: surveillance capitalism. It disaggregates the citizen into an expropriated bundle of wants and desires worked into algorithms and sold to the highest bidder for profit. This fracturing of the individual is the next step in the neoliberal process of undermining civil society and resistance to global neoliberal capitalism, precisely the ills AR aims to ameliorate.

How AR can address such complex challenges is not clear to me, but the strong voices of Akihiro Ogawa, Yoland Wadsworth, Shankar Sankaran, and now the impressive group of younger action researchers represented here fills me with an unaccustomed feeling: <u>optimism that a better future is still possible</u>!

Introduction: Learning, practicing, connecting, and passing along

Akihiro Ogawa

Abstract

This paper presents my action research (AR) journey – what led me to this paradigm, the key figures I have encountered, and my subsequent work over the past twenty years both in Japan and Australia. While developing my research program on civil society in contemporary democracies, I have committed to AR since my graduate days in the early 2000s. Based on my AR training under Davydd Greenwood at Cornell University, I have been conducting an AR project for almost two decades now in Tokyo, which started originally as a doctoral project. Since 2016, I have been coordinating a PhD seminar on AR at the University of Melbourne. I see this as me passing on my AR knowledge and experiences to the next generation of AR-minded scholars as well as connecting them to veteran AR scholars in Australia. Thus, I can clearly see a new research culture emerging from the new generation.

Key words: Encountering action research, PhD seminar, action research community, research paradigm

> **What is known about the topic?**
>
> Greenwood (1998) discusses action research in a university setting
>
> **What does this paper add?**
>
> - This article is an account of a life history of being an AR-minded scholar and practitioner within the context of the existing scholarship, based on the authors' own learning of AR in a university setting
>
> **Who will benefit from its content?**
>
> - AR seminar facilitators in a university setting
> - Students of AR
> - AR dissertation writers
>
> **What is the relevance to AL and AR scholars and practitioners?**
>
> - This article documents a style of seminar introducing AR at a doctoral level in a university setting.
> - This article illuminates connections between AR academics and local AR communities.
> - This article would be a practical guide for AR dissertation writers about how to proceed with AR projects.

Received November 2021   Reviewed July 2022   Published November 2022

Encountering Action Research

I first encountered Action Research (AR) in December 2000 at the Olin Library at Cornell University while struggling to define the direction of my graduate research. Frustrated with positivistic social sciences, I was looking for something different, such as knowledge proactively embedded in peoples' lives. As a researcher, I am not unemotional, objective, nor value neutral. I favour a socially engaged and democratic practice. AR has a way of changing things around in my life.

As I was looking through the titles of books, *Introduction to Action Research* (Greenwood and Levin, 1998) caught my eye. As I picked it up, I realized I was more interested in the subtitle – *Social Research for Social Change*. This was a life-changing moment, but it was no coincidence. My nerves were sharpened, trying to find something different and new. I was looking for something that I could materialize while exploring new knowledge. If I had not found this book on that snowy day in Ithaca, my life as a researcher would have been completely different. I might not have even chosen and continued my career as a researcher.

That day, I devoured the book, which provided me a new way of looking at things. There was a different academic world out there than the conventional social sciences in which I had been trained. The book discussed collaborative and co-generative knowledge production between researchers and collaborators, which continues in an (hopefully!) upward spiral curve, and eventually leads to solving problems that need to be addressed. I learned that AR is a cooperative knowledge process for social change through the researcher's strong commitment. After all, the book defined AR as 'social research carried out by a team that encompasses a professional action researcher and the members of an organization, community, or network ('stakeholders') who are seeking to improve the participants' situation' which leads to 'a more just, sustainable, or satisfying situation for the stakeholders' (Greenwood and Levin 1998, p. 1). This research strategy resonated with what I was looking for.

Ever since then, I have been developing a research program on civil society in contemporary democracies. My goal is to enhance grassroot access to the public policy agenda. This is based on my experience as a journalist before graduate school. I was a staff reporter for a Japanese wire service in the mid/late 1990s, where I covered Japanese policymaking dominantly led by elites – bureaucrats and politicians. How can ordinary people participate in the process? To answer this, I needed to know the relationship between the state and the individual where civil society could be an analytical lens. I define civil society as a public sphere that broadly refers to non-state institutions and associations that are critical to sustain modern democratic participation (Ogawa 2009, p. 2). The distinctive relationship, which reflects upon their culture and history, can be represented as civil society. However, my aim as a professional researcher was not to argue about what civil society *is* but to investigate what it *does*.

One of the authors of the book, Davydd Greenwood, was a professor of anthropology at Cornell University. I enrolled in his course, *Introduction to Action Research*, which was a weekly two-hour seminar throughout the semester (15 sessions). I learned from Greenwood theories of AR as well as methodologies for participatory planning and design, including search conferences, a collective process of inquiry that creates learning options for all participants (Greenwood 2007, p. 144), a technique originally created by Fred and Merrelyn Emery (see Emery and Emery 1974, for example). The course was led by students' presentations and discussions, and it enhanced our learning based on what we wanted to know and what we could offer (see my reflection in Ogawa 2006).

I transferred to anthropology from political science, where I was rescued by Greenwood from a life of positivism or researching theories only for the sake of it. Since then, AR has become the core of my work. In my first book, I wrote my mission as a professional researcher and defined my goal as follows: 'My ultimate objectives as an anthropologist in doing this type of research are to help empower ordinary people and to forward the democratization of society via the practice of action-oriented research' (Ogawa 2009, p. 19).

Base for my long-term practice

Since 2001, I have been practicing AR at a non-profit organization (NPO) promoting community-oriented lifelong learning in a place called Kawazoe (pseudonym) located in eastern Tokyo. I have documented the development of an action research project that I have been undertaking for two decades (Ogawa 2004, 2006, 2009, 2013, 2015, 2020, 2021). My field site, SLG (pseudonym), was established in September 2000 and disbanded in March 2018. It was part of Japan's newly institutionalized civic sector in the late 1990s following the surge of volunteerism after the Hanshin Awaji earthquake, which hit western Japan on January 17, 1995. The Japanese government took advantage of this sentiment and enacted a law supporting the establishment of NPOs by easing the rigid regulations on the incorporation of the third sector.

Among the NPOs, SLG was one of the largest community-oriented lifelong learning groups in terms of membership numbers and budget size. It was part of the government-organized nongovernmental organizations (GONGOs) that were created by the political process but operate quasi-independently of their establishing agencies, as well as of the organizations that implement government oversight of their economic or professional activity (Salamon and Sokolowski 2016, p. 1534). The municipal government supported SLG by providing generous funding to the tune of one billion yen (9 million USD) over the 18 years of its operation.

From 2001 to 2003, I worked at SLG as an unpaid secretariat staff member in charge of developing lifelong learning courses and coordinating volunteers. This was part of my fieldwork as a doctoral student. Since 2004, I have been involved as a volunteer that worked closely with the SLG's board of directors to initiate collaborative inquiry in search of breakthrough. Specifically, my project aimed to address the problems we faced at SLG to enhance the quality of organizational life. SLG members and I jointly repeated a cyclical process for problem-solving or co-generative knowledge production with our stakeholders: (1) identifying a problem, (2) gathering information, (3) analyzing the collected data, (4) planning for transformation, (5) taking action, and (6) interpreting the results to improve the quality of their social and organizational life through broad participation of the stakeholders.

Even though the SLG was disbanded in March 2018, the members/residents of Kawazoe have continued their learning activities. I have also continued to work with them even without any support from the government, as the people in my field site have continued developing their learning on local history and culture. In this way, my research went on in Kawazoe, exploring how learning is constructed and reconstructed after SLG (see Ogawa 2020a, 2020b, 2021). I have been collecting detailed accounts of people's expressed emotions, conflicts, examples of relationship building, and so on.

Consistent with the SLG project, I have also been developing another action-oriented social research in Japan since the catastrophic earthquake, the devastating tsunami, and the consequent radiation leakage caused by the subsequent meltdown at the Fukushima Daiichi nuclear power plant on March 11, 2011. I have been engaged in a decade-long ethnographic inquiry into Japanese civil society's reactions following this triple disaster which culminated in the book, *Antinuclear Citizens: Sustainability Policy and Grassroots Activism in Post-Fukushima Japan* (forthcoming). I developed a type of narrative, which I termed 'action narrative.' It illuminates my commitment as a researcher to document and co-create knowledge about the issues that citizens face, and jointly generate actions for social change. My narratives are socialized commitments for envisioning our future. I have documented ordinary people's actions as they survive and navigate a new reality in post-Fukushima Japan. The search for sustainability requires the development of collaborations in society that are grounded by voices arising primarily from civil society. These are the voices that challenge hegemonic ideology and practices, locally and globally. The dynamic interface between radiation-tainted environments and human societies is mediated by changes in actions that are carefully processed by civil society. The primary agents for change are 'antinuclear citizens' – conscientious citizens who envision a sustainable life in a nuclear-free society. Throughout my research, I address a key question: how have grassroots civic actions exploring sustainability shifted national and global agendas?

Bringing AR to Melbourne University

My academic affiliation has changed with the development of AR projects. After finishing a postdoctoral appointment at Harvard, I moved to Sweden to start a position at Stockholm University. I spent eight years there (2007-2015), primarily teaching area studies courses on Japan and Asia and was promoted from assistant professor to professor. During my last years at Stockholm, I started looking for a new opportunity to expand my research and teaching capacity. I regularly checked various academic job sites and found a chair position in Japanese Studies at the University of Melbourne. I applied in November 2014, visited the campus in February 2015, and received an offer shortly after.

When I accepted the offer in May 2015 and started in September 2015, I was asked by the Faculty of Arts to submit a proposal for a PhD elective subject. As I was told during my appointment that I did not have to restrict myself to Asian or Japanese studies, I was considering an interdisciplinary topic that could bring a new perspective to the university. I submitted two proposals in my areas of interest – Introduction to Action Research and Anthropology of Public Policy to the faculty. Eventually, they chose my proposal on AR, as the university was already offering a subject similar to policy analysis (Critical Approaches to Social Policy). I sent a simple proposal including subject title and overview or course description, along with teaching hours and assessment. These latter two were decided in line with other PhD subjects, so I could not make any changes. Here is a part of my proposal.

> Title: Introduction to Action Research Course description:
>
> Action Research (AR) combines new paradigms in social science research methods with a strong orientation toward democratic processes of social change. This interdisciplinary course focuses on how researchers and community members collaborate to conduct research on solving problems to improve people's lives. The main goal is to provide doctoral students with an understanding of useful theories, strategies of AR, an appreciation of the advantages and disadvantages of this strategy, and research skills needed to work as an AR researcher. The course also intends to introduce several specific AR projects. Toward the end of the course, students will be expected to design an AR project on a relevant topic. The primary course format reflects participatory commitment to co-teaching and co-

learning. Experts and practitioners will join discussions to broaden idea generation for our intellectual endeavor.[1]

Contact hours: Fortnightly, 2 hours x 6, total 12 hours

Assessment: One 500-word essay proposal[2] during the teaching period (20%) and a 2000-word essay[3] within four weeks of teaching completion (80%).

I was not informed what the faculty discussions were about when selecting my proposal or who chose it but I was informed by email that the faculty selected my proposal on *Introduction to Action Research* as a PhD elective subject which would start in 2016. I was excited for this new opportunity since I would be able to facilitate learning on AR, which is what I wanted for my academic career. I wanted to spread awareness about the action-oriented research strategy among emerging scholars and share my AR experiences. AR is a choice or a way of doing to develop research that is challenging and time consuming, but definitely worth trying and

---

1   This narrative was used for the course syllabus as well.

2   The writing assignment is a self-reflexive account. This is also part of the formative evaluation of the course development as well as students' thoughts on their performance informed by the AR literature. Through the paper, the students can react and integrate their thoughts and ideas on the proposed readings, their experience in the project, and the challenges they uncover. After submitting this paper, I organized a consultation with each student. Due to the university-wide requirements, PhD seminars only require a total of 2500 words for the assignment, and this paper requires 500 words. But my students wrote more, usually 1500-2000 words.

3   The writing assignment is another self-reflexive account. Students were asked to discuss their involvement throughout the semester from one of the angles listed below. Each student can grade his/her self and justify it. This assignment includes, but is not limited to, the areas below: (1) how student researchers engaged in the steps of AR, how they recorded their data and how they were a true reflection of what was studied; (2) how they challenged and tested their assumptions and interpretations of what was happening on a continual basis; (3) how they accessed different views, presenting confirming and contradictory interpretations; and (4) how these interpretations and analysis were grounded in academic theory and how this theory confirmed and challenged the analysis.

rewarding. I was very happy because I did not expect it to happen so soon.

In the planning, I tried to include all important points about AR over the six seminars. It was difficult, as the teaching hours for PhD electives at Melbourne were 12 hours, less than half of that at Cornell. Thus, I tried to effectively link my syllabus with the list of proposed reading on AR, which was available on my academic website (https://akiogawa.com/action-research/). This way, my students could also explore their research interests. My syllabus for the seminars included the following discussion questions:

Seminar 1: Introduction

o   Identifying offers and needs

Seminar 2: What is action research (AR)?

o   What is AR, and how does it differ from traditional social research?
o   What are the advantages and disadvantages of AR?
o   What are the goals of the AR process and product?
o   What are the key components of AR?
o   How does your research orientation align with this research strategy?
o   Why AR?

Seminar 3: Knowledge

o   What constitutes knowledge?
o   What does it mean to know something to make a change?
o   How can we gain new knowledge through AR?

Seminar 4: Positionality / Engagement

o   How does one's identity and social role affect research (specifically in AR)?
o   How does subjectivity affect research?

Seminar 5: Writing about AR[4]
- What does an AR dissertation look like?
- How do we write about AR?
- What are the important points to consider?

Seminar 6: Evaluating AR through research sharing[5]
- What does good AR look like?
- How do you justify your action-oriented research?
- How do we respond to institutional human subject reviews?

I proposed 'Discussion Starters and Facilitators' assignments. These exercises require rotating members of the class to create a short set of notes (5+ PowerPoint slides for a 15-minute presentation) based on recommended readings for each seminar. It is intended to help them organize their thoughts and prepare for class discussion. These exercises go beyond simple summaries of the readings; they are to reflect on how the ideas presented articulate with each other and with actual research projects. There were also added initiatives, such as additional readings from the

---

[4] The syllabus has been modified every year based on the previous years' experiences, in particular, the last two seminars. Instead of the topic on how to write AR, I invited Yoland Wadsworth for her talk, 'Practicing Action Research' in 2017 and 2019. In 2017 and 2018 I invited Chair of the Faculty of Arts Human Ethics Advisory Group in the Faculty of Arts to discuss ethics in AR. Starting 2020, I changed the seminar agenda. Seminar 5 talks about collaborative inquiry. Our discussion focuses on case studies in order to look at 'How does a collaborative inquiry work?' and 'How do researchers overcome tensions with collaborators?'

[5] From 2017 to 2020 the final seminar was used for sharing ongoing projects by all students. In 2021 I changed the agenda to Research Design. This change encouraged students to spend possibly more time on reflection and planning for their PhD projects. Furthermore, I invited some (mostly two or three) students from the seminar to share their ongoing projects. Wadsworth joined this seminar as an AR veteran. I made these changes as I saw differences of understanding about AR among the students. Some have clear planning on their AR-oriented projects. Meanwhile, some are still in the early stage of developing the projects.

list of proposed readings and/or one's own way to enhance the group's understanding of AR. Students are expected to continue their discussion on the course website after class.

In July 2016, almost one year after arriving in Melbourne, I started teaching AR, or more specifically, facilitating it with eight doctoral students. Since then, I have been coordinating this seminar as a PhD elective every second semester. On the first day, I emphasized in my introduction that the primary course format should reflect the participatory commitment to co-teaching and co-learning, while conveying the message that AR is challenging and time consuming, but meaningful and rewarding. The first task I proposed was a search conference to identify needs and offers (what students need to know and what they can offer to this course). I continue this task throughout my facilitation. Here, I share a list of offers and needs from the newest cohort in 2021 with their permission.

Offers

- Experience in developing multidisciplinary international creative projects in communication, film, and media in different countries (India, Mexico, China, and the Netherlands) and how this knowledge can contribute to the AR methodology.
- Resources that I can share with my experiences and learning from planning and implementing AR.
- Current perspectives to embed AR strategy and cohort knowledge exchange.
- My learned experience negotiating with stakeholders within a (sometimes fraught) conservation field.
- My background in journalism, development communication, stakeholder engagement, and human rights practice and research.
- Stakeholder management and negotiation skills (business development and management experience), international relations background

- Cultural heritage conservation approaches such as object biography, decision-making, scientific analysis, interdisciplinary collaboration, and so on.
- I have rich fieldwork experience (five different ethnic minority areas in China) in face-to-face interviews with a large group (about 100 people) to share.
- Building communication with stakeholders from the 'art industry' (museums, galleries, art fairs, artists, etc.)

Needs

- How to approach, involve, and prompt the participation of institutions, authorities, professionals, and the general public on an international scale.
- How to share ideas and to ask critical questions.
- Complex field experience in the community.
- To understand how AR can fit with conservation practices, and how I can continue to consult with stakeholders in the conservation of artefacts which can be difficult based on the age of the object and how many/diverse stakeholders are involved.
- How I can better engage and empower my thesis stakeholders/ sources through AR
- Likeminded researchers to help develop an understanding of AR through critical discussion and discourse.
- My focus is on 80s street art. I would like to find out the significance of street art in Melbourne, maybe through communities and interviews.
- To learn more about the classic cases of AR in social research areas and how it fits into my thesis (poverty alleviation practices).
- General understanding of AR strategies and how they can be applied to my thesis topic

The students shared this list of offers and needs as reflection points on a thread in the discussion forum of the course website. Eventually, they resumed this conversation on WhatsApp to explore further collaboration in their PhD journey in AR.

Over the past six years from 2016 to 2021, 40 students have enrolled in this course. In 2021, 10 students enrolled, which was a record number thus far. There were many action-minded doctoral students from all departments of the university, including arts (anthropology, sociology, theatre arts, philosophy, political science, conservation studies, indigenous studies, linguistics, creative writing, and translation studies), education, and business/economics. Along with the regular first-year PhD students, I also had a couple of auditing students (mostly second years) who had finished the candidate examination. They could not take my seminar in the first year, but they joined me later on when they could. I also added a paragraph below in the course description after 2020.

> Toward the end of the course, students are expected to design an AR project on a relevant topic. Since the start of 2016, many AR-oriented dissertations have been produced, which has contributed to the generation of a new research culture at this university. Previous students' projects have included cultural heritage management, human resource strategy, museum exhibit creation, youth theatre, and social movements' engagement.

As many students are interested in AR, I have invited former students to the ongoing seminars since 2017 and asked them to share their project progress with the current cohort. I foresaw a win-win situation: the seminar students could witness and be inspired by the actual development of AR projects, while the former students might encounter some reflection points by sharing their ongoing projects. In 2020, the faculty added 'action research' as a box one can check in the research ethics application. Students need to justify the research. It defines AR like this:

> Action research is often community - or organisation- based and is carried out in the field. This approach involves

testing ideas in practice as a means of improving social, economic or environmental conditions and increasing knowledge. Action research proceeds in a spiral of steps consisting of planning, action, and evaluation. It provides a basis for further planning of critically informed action. This method includes design and implementation research and rapid appraisal research.

I realized that the seminar participants and I have committed to creating a new research culture at this university. This research strategy is likely going to increase in prevalence, and mark a significant presence among our doctoral students.

Connecting to the AR community in Australia

I also tried to connect (and re-connect) with the AR community in Australia, to ideally connect my students to the AR community. Even before I came to Melbourne, I have known since my graduate days that Australia was very active in AR. I realized in retrospect that many researchers cited in my papers were from Australia. When I took Greenwood's course at Cornell in the spring semester of 2001, I remember a group of my fellow students attending an AR conference in Brisbane; however, I could not make it because of an exam. Handbooks and encyclopaedias on AR have included many names of Australian-based researchers and practitioners among their authors (e.g., Bradbury 2015; Coghlan and Brydon-Miller 2014).

One central figure in Australian action research is Bob Dick. He is an independent action-oriented practitioner based in Brisbane on the east coast. He is retired but he remains a central figure in AR. In fact, I have been browsing his AR website (http://www.aral.com.au/) since I was a graduate student. It is the only site where we can obtain comprehensive knowledge on AR, with all the necessary information about the theory and practice in a compact format. The Action Learning, Action Research Association (ALARA), a practice-based academic organization for AR in Australia, was founded by the Queensland network of which Bob Dick was a member. Under the leadership of Ortrun Zuber-Skerritt and other colleagues from business and higher education fields, ALARA became one of the two main international action-research networks, with ALARA based in Australia and CARN (originally Classroom AR Network) based in the UK.

The first person I met after my arrival in Australia was Shankar Sankaran (currently teaching at the University Technology of Sydney), whom I previously met at Cornell in October 2005. I had just finished my Ph.D. and moved to Harvard to start a postdoctoral fellowship. I received an email from Greenwood informing me of Sankaran's visit to Cornell. My wife and I drove four hours from Boston to Ithaca with our one-month-old daughter in the back seat to attend his lecture. He was teaching at Southern Cross University at that time. He has extensive experience in the field of project management at the Asian subsidiary of Yokogawa Electric Corporation, and has been actively consulting on corporate organizational culture which employs AR. In October 2015, we met again in Sydney.

After this meeting, Shankar sent an email to his colleagues and friends in the field of AR in Australia. I received a series of warm welcome emails, one of which was from Susan Goff, a social ecologist and an AR practitioner, saying 'You are participating in a development in Australian academic life that we have all been working towards for decades, and it is interesting that it should finally arrive in the form that it has. Perhaps your appointment is a sign of things changing for the better at long last'.[6] I also received an email note from Yoland Wadsworth, a name mentioned by Greenwood a couple of times during the graduate seminar. I did not know that she was Melbourne-based. I only knew from her works (Wadsworth 1984, 1997) that she was a key figure in Australian AR from a health, community, and human services practice perspective for the previous 40 years. She, like Bob Dick, is a retired researcher, consultant and an adjunct professor at RMIT University. Yoland immediately suggested that we meet in person,

---

6   Personal communication on October 5, 2015.

but I unfortunately had a research trip to Japan. Six months after the first email in May 2016, she delivered a copy of her book *Building in Research and Evaluation: Human Inquiry for Living Systems* (Wadsworth 2011) to me as a gift along with a letter to my office in my absence.

The letter informed me about the upcoming annual general meeting of the AR Issues Association (ARIA) in November and asked if I would be available to attend. ARIA is a not-for-profit community association that has been a longstanding local home for AR in Australia, as well as a community of practice for around thirty 'epistemological fellow travelling companions' in Melbourne and Victoria for the past 30 years. It is also a provider of resources to assist the co-inquiry processes of others worldwide according to her homepage (https://yolandwadsworth.net.au/, last accessed on September 1, 2021). However, I had not heard of this association before, and I wanted to know more about it.

In October 2016, I finally met Yoland in person and had a long conversation on her AR experiences and its history in Australia. I learned that she founded ARIA in April 1988, along with a group of fellow AR researchers to help people solve their own social problems in collaboration with relevant stakeholders. It started with 25 members. Each member wanted to actively help improve the conditions of those deemed (in the language of the 1980s) the socially disadvantaged. People who worked on their own AR with ARIA support or resourcing decided how they wanted to live their lives by participating in decision-making, thus maximizing their self-management. Through a participatory approach, the members firmly believed in the importance of an inquisitive mind and a critical reference group (or critical inquiry group). To connect my seminar participants with ARIA, Ai Ming Chow and Ed Hyatt from the 2016 seminar (both contributors to this issue) joined the annual general meeting in November 2016, where they were warmly welcomed by Melbourne-based scholars and practitioners. I heard later that both enjoyed hearing about the various projects going on in the community and were considering incorporating AR into their research. In 2017, I participated in the annual general meeting of the association and encountered the rich AR tradition in Melbourne. I became a member of ARIA and attended the 30th anniversary dinner in April 2018, which was held at an old Japanese restaurant in Melbourne's Chinatown, the same restaurant where they held their inaugural meeting in 1988. During the dinner, we listened to everyone's interesting stories on 'why were we still a member of AR Issues Association?' Lesley Hoatson, Jacques Boulet, Saide Gray, Bill Genat, Daryl Taylor, and Linette Hawkins, prominent Australian AR researchers, attended this meeting along with Yoland Wadsworth to celebrate. In 2017, I invited Yoland to share her AR experiences with my PhD seminar students. In 2019, she provided a Public Lecture for National Social Sciences Week jointly with my PhD seminar at the University of Melbourne at which she spoke about her career-culminating generation of a new transdisciplinary full cycle social science meta-epistemology, drawn from her four decades of practice-based participatory action research. Active interactions beyond generations have started.

I was also connected to another important community researcher, Ernest Stringer, who teaches at Curtin University in Perth on the west coast of Australia. Stringer has published several books on AR (e.g., Stringer and Aragón 2020). From the mid-1980s, based at Curtin's Center for Aboriginal Studies, he has had an immense amount of experience working with Aboriginal people and schools. In November 2016, while I was teaching my first AR course, I sent an email to Stringer to connect my students to him for potential collaborations and other benefits. In the email conversation, I came to know that Stringer was coming to Melbourne in March 2017. I organized a talk to share his stories with my students who were just starting out on their research journeys. Yoland Wadsworth also joined us after the talk, which led to a reconnection between two important AR scholars in Australia.

Finally, the members of ARIA also established the Action Research Centre in the city center of Melbourne. As a non-profit organization, the center managed many projects (with funding from the Victorian state government) together with the government and other community-based organizations for a total of seventeen years. They worked with community health centers, social welfare services, community groups, self-help groups, psychiatric and general public hospitals, and other organizations that needed co-inquiry with critical stakeholders. They held numerous workshops, forums and conferences to promote AR, action evaluation and systemic dialogue, and actively created opportunities for participation in the policy and evaluation processes of government services. It provided a place where social welfare service providers and service users could together, search for better ways to provide services and test new strategies in practice based off each other's experiences.

Wadsworth noted that she had been facilitating dialogue directly between the parties involved over a period of time, rather than merely act as a one-off, one-way messenger to a single commissioning party. Service improvement would emerge from following the evaluative processes of AR, leading to the next evaluation, and efforts to improve the situation may thus continue. The Action Research Issues Centre was a 'small center with a big agenda' that worked with thousands of people over the years. In the 1990s, their work received international acclaim, while Australia's best-selling book on action evaluation methods, *Everyday Evaluation on the Run* (Wadsworth 1997), was produced from a state government social justice strategy-funded project. It remains in print till this very day. ARIA members were also subsequently involved with the establishment of a research unit on AR at Victoria University in 1999 for two years. Yoland was then invited to Swinburne University of Technology in 2001 to develop and seek funding for a national Action Research Program which included research, consultancy, and an AR postgraduate course. This was all a first for Australia. In 2005, the graduate Collaborative and AR course was offered. However, the program was abruptly closed in 2006 due to structural changes within the university. Despite being fully externally funded, a well-documented demand that was growing by the day, and their excellent accreditations, new management from the laboratory sciences unilaterally deemed they saw 'no need' for action research.

Searching for a new research paradigm

What I envisioned in my AR seminar resonates with what Wadsworth and members of ARIA envisioned – the search for a new research paradigm. I believe that AR lies at the heart of that paradigm. I have been looking for a counterpoint to positivist research, which seeks absolute objectivity and is oriented toward value-free research without any value judgment by the researcher. I firmly believe that there can be another research paradigm in which the researcher is oriented toward becoming deeply involved in improving people's lives. By doing so, I argue that researchers would be able to better understand the problems ordinary people face in their daily lives. Research that does not deeply involve the people being studied is like removing the 'social' from social research.

While I am free to pursue AR in Australia, the long-term journey of AR is still in progress. When I was a graduate student in the early 2000s, I remember how difficult it was to convince the traditional and dominant positivist social researchers of the real appeal of AR. I can only imagine how bad the situation was in the 1970s, 1980s, and the 1990s when Wadsworth and other members of ARIA were most active. While the theories of AR had to be strong for the pioneer generation to mount arguments for it, AR was still not as institutionalized or as well organized as it is now. They certainly must have felt more resistance from their employers and colleagues than I had experienced.

I tell my students that academic scholarship has various kinds of knowledge production paradigms, and that each scholar can choose one. For instance, using Lincoln et al. (2011), in the beginning of the seminar, I showed five knowledge paradigms: positivism, post-positivism, critical theory, constructivism (or interpretivist), and participatory. I encourage my students to identify their knowledge paradigm or the most comfortable one to explain what they argue. This is a starting point for early academic researchers to carefully consider.

Each paradigm has a different ontology (or worldviews and assumptions in which researchers operate in their search for new knowledge), epistemology (or process of thinking, the relationship between what we know and what we see), and methodology (or the process of how we seek new knowledge). Furthermore, the nature of knowledge or how researchers view the knowledge generated through inquiry research is different. AR researchers believe knowledge is dialectically created, context-dependent, and provisional (Greenwood and Levin 2007). Positivists believe the hypothesis is verified; post-positivists believe there is a current single truth, but many have multiple hidden values and variables that prevent us from ever fully knowing the answer; critical theorists believe knowledge is viewed as subjective, emancipatory, and productive of fundamental change; while constructivists believe the constructed meanings of actors are the foundation of knowledge (see Lincoln et al. 2011, p. 106).

There is no superior or inferior nor right or wrong paradigm. The choice is primarily based on the ways in which each scholar achieves academic rigor and what kinds of scholarship s/he aims to establish. I tell my students that it is important to find the right audience when presenting their work if they want to have fruitful discussions. For instance, AR researchers working on a participatory paradigm do not employ the same language and conceptualization of reality as scholars in positivism do. There is no point in continuing such conversations. I have experienced this since my graduate days. Instead, each scholar needs to respect and understand the different modes of scholarship production. The knowledge that AR pursues is not knowledge for the sake of knowledge, but to improve the quality of life in a dialogical approach. I have learned a lot from Greenwood and other veteran AR researchers and practitioners since my doctoral work, after which I have been diligently practicing AR. Now, I pass my AR knowledge and experiences to students who have started their intellectual journeys of exploring something tangible and most meaningful to society.

Papers ahead

AR is emerging as a solid presence in academic circles. AR researchers are pushing a way of knowledge production forward, and testing where the boundaries lie in the academic world. Following this introduction – and the response of my AR colleague Professor Yoland Wadsworth – this special issue includes four papers from dissertation writers who are forthrightly challenging the conventional mode of knowledge production. The authors include Akina Mikami, Abbie Trott, Ai Ming Chow, and Ed Hyatt, who all took my PhD seminar at the University of Melbourne. The papers were originally presented in a two-day workshop for AR dissertation writers, which was scheduled for November 2020. Professors Davydd Greeenwood and Yolamd Wardsworth both attended the workshop. In the process of creating this issue, all writers also engaged in co-writing to form a reflexive account of their AR experiences in PhD research.

References

Bradbury, H,201, *The Sage Handbook of Action Research* (3rd edn), Sage Publications Ltd, Thousand Oaks, CA.

Coghlan, D & Brydon-Miller, M 2014, *The Sage encyclopedia of action research*, Sage Publications Ltd, Thousand Oaks, CA.

Emery, F & Emery, M 1974, 'Participative design: Work and community life' in Emery, M (Ed), *Participative design for participative democracy*. Centre for

Continuing Education, Australian National University, Canberra, pp. 100–122.

Greenwood, Davydd J & Levin, M 1998, *Introduction to action research: Social Research for Social Change* (1st edn), Sage Publications Ltd, Thousand Oaks, CA.

Greenwood, Davydd J & Levin, M 2007, *Introduction to action research: Social research for social change* (2nd edn), Sage Publications Ltd, THosand Oaks, CA>.

Lincoln, YS, Lynham, SA & Guba, EG 2011, 'Paradigmatic Controversies, Contradictions, and Emerging Confluences, Revisited', in Denzin, NK & Lincoln, YS (Eds) *The Sage handbook of qualitative research*, Sage Publications Ltd, Thousand Oaks, CA., pp. 97–128.

Ogawa, A 2004, *The failure of civil society?: An ethnography of NPOs and the state in contemporary Japan*, Cornell University Department of Anthropology, PhD dissertation.

Ogawa, A (2006) 'Initiating change: Doing action research in Japan', in Gardner, A & Hoffman, DM (Eds) *Dispatches from the field: Neophyte ethnographers in a changing world*, Waveland Press, Long Grove, pp. 207–221.

Ogawa A 2009, *The failure of civil society?: The third sector and the state in contemporary Japan*, State University of New York Press, Albany.

Ogawa A 2013, 'Lifelong learning in Tokyo: A satisfying engagement with action research in Japan', *Anthropology in Action*, vol. 20, no. 2, pp. 46–57.

Ogawa A 2015, *Lifelong learning in neoliberal Japan: Risk, knowledge, and community*, State University of New York Press, Albany.

Ogawa A 2020a, 'Japanese NPOs and the state re-examined: Reflections eighteen years on', in Chiavacci, D, Obinger, J, & Grano, S (Eds), *Civil society and the state in democratic east Asia: Between entanglement and contention in post-high growth*, Amsterdam University Press, Amsterdam, pp. 219–238.

Ogawa A 2020b, 'Kaigai kara mita Nihon no shakai kyōiku shōgai gakushū ~ korekara don'na dezain kufū ga hitsuyō ka —'sumida gakushū gāden' no jirei wo hikinagara' [Social education and lifelong learning in Japan from an overseas perspective ~What kind of designs and innovations will be needed in the future – Drawing an example from Sumida Gakushu Garden], *Shakai Kyōiku* vol. 889, pp. 10–13, Nippon-Seinenkan, Tokya. [In Japanese]

Ogawa A 2021, 'Civil society in Japan', in Pekkanen, R & Pekkanen, S (Eds), *Oxford handbook of Japanese politics*, Oxford University Press, New York, pp. 299–316.

Ogawa A (Forthcoming) *Antinuclear Citizens: Sustainability Policy and Grassroots Activism in Post-Fukushima Japan*, Stanford University Press, Stanford.

Salamon, LM & Sokolowski, SW 2016, 'Beyond nonprofits: Re-conceptualizing the third Sector', *VOLUNTAS*, vol. 27, no. 4, pp. 1515–1545.

Stringer, ET & Aragón, AO 2020, *Action research* (5th edn),Sage Publications Ltd, Thousand Oaks.

Wadsworth, Y 1984, *Do it yourself social research*, Allen & Unwin, St Leonards.

Wadsworth, Y 1997, *Everyday evaluation on the run*, Allen & Unwin, St Leonards.

Wadsworth, Y 2011, *Building in research and evaluation: Human inquiry for living systems*, Left Coast Press, Walnut Creek.

Biography

Akihiro Ogawa is Professor of Japanese Studies at the Asia Institute at the University of Melbourne. His major research interest is contemporary Japan and Asia, focusing on civil society and politics. His publications include the award-winning book, *The Failure of Civil Society?: The Third Sector and the State in Contemporary Japan* (SUNY, 2009), and edited volumes such as *Routledge Handbook of Civil Society in Asia* (2017) and *Transnational Civil Society in Asia: The Potential of Grassroots Regionalization* (2021). His new book – *Antinuclear Citizens: Sustainability Policy and Grassroots Activism in Post-Fukushima Japan* – is forthcoming in June 2023 from Stanford University Press. He can be reached at akihiro.ogawa@unimelb.edu.au.

A 'veteran reflects' – A personal biographic response to Akihiro Ogawa's Introduction

Yoland Wadsworth

Received February 2022   Reviewed August 2022   Published November 2022

Introduction

I offer my response to Akihiro's biographic account in the hope it may have something to say about the differences and the parallels in our different eras of lived experience with Action Research (AR) and the 'passing of the baton' between generations.

For example, it has been interesting to reflect on Akihiro's account of first encountering AR in December 2000 in the Olin Library at Cornell University. Only a few minutes' walk away from there, in another Cornell University Library Hall, I had something of a culminating moment when I presented to a public and multi-disciplinary academic audience in 2014 the results of my previous 42 years of AR practice – a transdisciplinary theory of Action Research as a meta-epistemological process in which inquiring is identified as the dynamic of all living systems (Wadsworth 2008a, 2008b, 2020).

Akihiro has been able to characterize his AR journey as having started with (1) *Learning* at Cornell with Davydd Greenwood; then having applied it (2) in his *Practising* in Japan with a lengthy engagement with civil society activists, building on his earlier political science and journalism; then (3) *Connecting* with AR 'veterans' in Australia, and finally (4) *Passing Along* all he has learned in his well-grounded AR course at the University of Melbourne since 2016. In contrast, as one of the generation that 'made the road by walking it' here in Australia, the order of my encounter with AR is somewhat different.

With no AR books readily available in libraries to learn from, nor even the language of AR in Melbourne in the early 1970s (at least not that I knew of), I initially found myself conducting mainstream social science in public health services only to find myself soon needing to (1) *Connect* with others who were having similar experiences with the painful and frustrating inappropriateness of orthodox positivism in an era of immense social change. Then we were (2) *Practising* (literally!) through trial and error in our various research projects and soon creating more formal groups and networks in which to actively share the puzzles of what we were doing and how to overcome them, and then (3) *Passing it along* to the many others who were being pressed into carrying out 'do it yourself research', and later 'everyday evaluation on the run', whether as practitioners, service-users or communities expressing needs for the new services we were researching-and-developing (R&D). This included responding to the need for a 'new paradigm' literature, and some of us – like Davydd Greenwood, John Gaventa, Patricia Maguire, Jennifer Greene, Yvonna Lincoln and Egon Guba in the USA; Peter Reason and Judi Marshall in the UK; Stephen Kemmis and Robin McTaggart, Ortrun Zuber-Skerritt and myself in Australia; Anisur Rahman in Bangladesh; Rajesh Tandon in India, and Paulo Freire, Orlando Fals Borda and Daniel Selener in South America, among many others – were beginning to write the books that later searchers such as Akihiro would be able to find in a library. But at the time, it was all very much 'live in action' or 'on the hoof' as Lynton Brown later put it in his 1988 book for the state Education Department on Group Self Evaluation.

Since then, we continued (4) *Learning* more deeply how – as I would now put it – to 'inquire for life', and in my work so that there might be better 'built in' inquiring to all aspects of not only health and community services, but of human and more-than-human life per se.

Pondering also on Akihiro's reference to my being one of the 'veterans' of AR at first made me feel a bit like an 'old soldier'!

Especially when I like to think of myself as having only rather recently embarked on illuminating my transdisciplinary living systems meta-epistemology. Plus, I've also always been pretty dedicated to peaceful co-inquiry through dialogue rather than militaristic means for resolving human differences! But the etymology tells me that a veteran is indeed an 'old soldier', and, although the US Army defines that as being as little as six months' active service, on reflection, I think I *have* been 'soldiering away' for 42 years of 'active service' in the epistemological 'paradigm wars'. I certainly have the scars to show for it!

Connecting

From the time of my employment outside academe as the Victorian Health Department's first Research Sociologist in 1972, I had the same growing sense of frustration with positivism's dominance as did Akihiro all those years later.

I had discovered, for example, through research I conducted for the Victorian Government's cross-portfolio Consultative Council on Pre-School Child Development, the profoundly different meanings available from using a conventional state-wide questionnaire survey with Victorian parents of young children (analyzed by the Statistical Package for the Social Sciences [SPSS] for me by a white-coated mathematician on an immense mainframe computer down the corridor) – compared to the meanings available from questioning the same parents in local personal face to face interviews. I wrote at the time the former had felt like 'reading braille through a doona' and the latter began to bring things to life.

The survey told me 'how many thought X versus Y', but it was the interviews that gave me rich descriptive and insightful explanatory 'knowledge proactively embedded in peoples' lives' (in Akihiro's words), not only about the how and why of the 'x and y' but also regarding other factors that may have been more significant in the complex lives of young parents and their families living through a period of immense social change.

I recall in those days (remarkable now in a time when 'quantitative' and 'qualitative' are so taken-for-granted as multi-method approaches) that I had to argue strenuously even to be allowed a 10% interview sample since the Permanent Head of the Health Department thought an 'objective questionnaire survey' should surely suffice. Fortunately, I worked directly for the Chief Health Officer, who was a quietly wise and deeply philosophical man, and a path-breaker in his own time (he had helped establish the Australian Humanist Society), and he was acutely aware that new times demanded new approaches.

I, myself, had just been through a social science course in research methods that had involved us marching over to the other side of the campus to the Mathematics Department to learn how to calculate Chi-square tests of significance by hand (truly!), so I had more than an inkling of where the Permanent Head was 'coming from'. Social researchers were still very much expected to 'wear the white coat' of clinical lab scientists. However, cracks were appearing in the hypothetico-deductive paradigm's dominance as it was not able to offer ways of effectively understanding what was happening in the world in order to develop new responses to increasingly gaping holes in a wide range of health, education and human services, mostly operating in a highly standardized way on the same models that had been developed for the conditions of life in the early 1900s or earlier

Fortunately, my classical Monash University sociology education (British empiricism, American pragmatism and European theory) had also taught about the Chicago School of urban sociology, and theories of change as well as a critique of structural functionalism. We read George Herbert Mead and Robert E. Park and their ethnographic methods of field research, and about symbolic interactionism, interpretivism, phenomenology, critical constructivism, and people such as Howard S. Becker, Erving Goffman, C Wright Mills, Peter Berger, William Foote Whyte, Alvin Gouldner, Paulo Freire and Anthony Giddens. This equipped me to argue for more engaged methods and take part in the 'paradigm wars' along with a larger group of practice-based change-oriented social researchers outside academe. Later, my sociology colleague Lucinda Aberdeen and I consolidated this non-academic network as a lively monthly seminar group titled MERGe (Melbourne Evaluation Social Research Group, etc.) with its own journal and membership directory.

Practising

In 1974 when I was living in London, I met up with some urban researchers at the London Polytechnic who were using 'participatory action research' (PAR). I had the sense that *this* was more like what I was looking for, so when I received an invitation from the Chief Health Officer to return to Australia to evaluate the pilot local level integrated model of services that I'd helped sketch (literally) for the government review, I said yes 'but only if I could use Participatory Action Research'. The CHO agreed, although neither he nor I really knew what that meant, but we spent the next three years finding out! It led to an ethnographic local community needs study with me moving from being a 'participant-observer' to an 'observing-participant' and inventing new methods like kitchen table conversations to co-inquire with people 'in the middle of life' as and where it was being lived. This was followed by a services evaluation that saw me moving even further from being 'the researcher' to being a 'facilitator of emergent co-inquiry' (Wadsworth 2011a, originally 1984). This deeper work, developing a comprehensive services framework at the local level, was still being used in municipal government policy-making twenty-five years later.

It was very much like the subtitle of my colleagues Davydd Greenwood and Morten Levin's 1998 book, which so attracted Akihiro – '*Social Research for Social Change*' – although these terms were not acceptable currency at the time when to speak about researching for 'social change' was tantamount to admitting to the 'unscientific bias' of *valuing* change or valuing *anything* really. 'Subjective' opinions and values were what 'subjects' had. Neutral researchers were value-free and 'objective'. However, we soon learned the fallacy in that, and became skilled at arguing the logical necessity of values to guide any successful inquiry per se.

Passing It Along

Soon action research books were becoming more readily available from overseas, so our 'small centre with a big agenda', the Action Research Issues Centre and its incorporated Action Research Issues Association (ARIA) became a publisher in 1986 and began importing and selling, then writing and publishing a small methodology literature list of its own. We then embarked on 17 years of supplying action research consultancy, workshops, guest lectures and publications, and a growing swathe of formal networks in Victoria – Friends of Participatory Action Research (FOPAR), Teachers of Participatory Action Research (TOPAR), Researchers in Community Health (RICH), and Systemic Participatory Inclusive Research Action Learning (SPIRAL), with accompanying journals, newsletters, seminar programs and membership directories. We later connected with comparable initiatives interstate, nationally, and eventually, internationally[1], but always retained our local focus.

In the 1984 *Do-It-Yourself Social Research* book that I wrote to demystify the process after ten years of applied research experience, although I describe (and depicted on the cover) a recognizable cycle of action research, I called it 'social research' because it seemed to me the logic of *all* social inquiry, but particularly for human service improvement, required a participatory action model as the only one that made sense if you really wanted to find out with people what their conditions were really like and how they lived and worked under those circumstances including with their own purposes. I particularly learned how such values-driven purposes could effectively shape

---

1 It is beyond the scope of this paper to document the extensive network of AR colleague-friends that ensued from these wider collaborations, but Wadsworth (2014) documents the main organisers, keynote speakers and themes of all the ALARPM & PAR World Congresses up till then.

the inquiry in the hard test of practice precisely because those participating in the research: parents, community residents, professionals, administrators and funders who all, more or less, shared those deep values – needed the services to respond accurately to the critical inquiry group, to align effectively with those deep values.

From 1978 to 1984, I returned to Monash University to make sense methodologically of nearly ten years' challenging co-inquiry efforts and to get to the bottom of the philosophy of science and sociology of knowledge literature for a PhD in which I wanted to identify what constituted both a methodologically-sound and practically-useful approach. *Do-It-Yourself Social Research* was the concrete result, and the book was bound into my 1984 thesis; then, unbelievably, became Australia's best-selling book on social research, never out of print to this day. (I love pointing out to students that a fellow post-graduate student wrote it *and* one who was working outside academe!)

Soon after, Stephen Kemmis and Robin McTaggart published their 1988 *Action Research Planner* drawing on their work in schools and the education sector. Like ARIA's community of practice in health, human services and community-based practice, the Deakin University Education Faculty also formed a powerful community of practice in schools and with policy researchers in the state Education Department to bring together theory and practice.

Later I wrote *Everyday Evaluation on The Run* (published initially by ARIA but now, like *Do It Yourself Social Research*, Routledge, also a surprising best-seller never out of print since 1991) to demystify and describe the process of a practice-based evaluation culture at a time evaluation was rapidly emerging in the 1980s and 1990s. I think the necessity of values clarification in evaluation dealt a death blow to positivism's 'value-free science' and simultaneously made it obvious why all stakeholders needed to actively participate to address the value-driven purposes of the critical stake-owner group. In the 1990s, ARIA produced some popular papers *'What is participatory action research?'*, *'What is feminist participatory action research?*, and *'How can professionals help groups do their own participatory action research?'* – additionally resulting from those nearly two decades of applied PAR and evaluation.

Learning More Deeply

Like Akihiro, I have been fortunate to have had in-depth field sites in which to work over longer periods of time that informed these writings. In the 1970s, mine were with parents and families in local communities with connections to large state bureaucracies; in the 1980s, with community health, public health policy and disability deinstitutionalization; in the 1990s with acute psychiatric hospital service users and providers, community mental health and general hospital services, and in the 2000s with tertiary education, welfare and health promotion services and their users.

Later, this more in-depth style of work was curtailed by the government adopting the global managerialist ideology of doing endlessly 'more for less', instituting harsh cuts to the public sector in the name of quite short-sighted 'efficiency'. While so many of us and our projects suffered either harsh ends (often coming with a change of government) or 'death by a thousand cuts' (by governments in office), those 30 years of AR and PAR and community development-based processes and other new approaches arising from the forthright creativity of so many communities-of-interest, I think paved the way for the resurgence of current interest in co-design and deliberative processes (albeit these latter are often coming from a highly empowered and professionalized 'top down' rather than entrusting the navigation of the inquiry process to its less empowered critical inquiry groups).

So, interestingly, Akihiro brought action research teaching back to Melbourne at a time when it felt like AR here in Victoria and elsewhere had been quite badly diminished by neoliberalism. We wanted to support his interdisciplinary PhD course to see AR flourish again, especially as he was bridging from Davydd Greenwood and the previous era of PAR flourishing at Cornell University.

Akihiro notes now that AR is likely to increase in prevalence, making a significant presence among the doctoral students, and that the teaching of AR at the University of Melbourne is beginning to institutionalize as a culture. Of course, as Akihiro also notes, the jury is out on the larger project of ending the paradigm wars, which currently seem to have re-erupted globally in the form of a renewed insistence on science, technology, engineering and mathematics (STEM) and an associated onslaught against the humanities, history, arts, social sciences, history and other life approaches (HASS). Ironically this turn to STEM is often in the name of gender equality and girls' increased access to STEM subjects when it was women researchers who were so key to popularising the practice of qualitative methods in HASS. It remains telling that the logically matching campaign to encourage more boys into HASS has yet to materialize. Nevertheless, we see tendrils of life growing between the two fields, particularly as alarm grows at the dangerous new imbalance forming.

Akihiro describes his skilful, practical accommodation of the differing philosophies of science favoured by his students. In a parallel but different vein, the 'new, new paradigm' that I developed (Wadsworth 2008a, 2008b, 2010) maps these as 'inquiry preferences' around a meta-epistemological living systems AR inquiry cycle 'backbone' – at the level of the personal psychological in methods choices, and 'writ large' as the differing philosophies of science at the level of the social and sociological.

Applying this new 'mental architecture' or framework enables seeing the old 'paradigm wars' as between *one half* of that meta-epistemological AR cycle (the rational planning hypothetico-deductive thinking-observing, measuring, experimental action, 'how things are' half) – versus *the other half* (of inductive-feeling-reflecting, 'how things could be' constructivist abductive-theory-generating). The field framed these mostly as positivist vs anti- or post-positivist, or as quantitative vs qualitative, or empiricist vs interpretivist / constructivist / constructionist, and an uneasy truce was eventually reached that more or less allowed for coexistence.

In contrast, the 'new, new paradigm' might be seen, not as additive, but as an *integrative* project of seeing the two halves joined in one cyclic dynamic. I find myself repeatedly chanting, regarding the many inquiry capabilities that 'get us round the cycle', that: 'we all do all, we all *can* do all, we all *must* do all to traverse the cycle to make our lives – *and* we have preferences for some of these inquiry capabilities and not others' (Wadsworth 2015). These personality preferences seem to be what makes the difference for conflict, as these differences are currently being mistaken all round the world – not as our necessary epistemic diversity and 'gifts differing' or 'the inquiry capabilities of living systems' – but as 'our' preferences being the best, and 'the other's' being unnecessary, inferior, wrong, misguided, or dangerous.

My own life's work has in this way led me to seeing this 'new new paradigm' *not as replacing* the old paradigm but *bringing it back to life*, and all its elements into the right relationship or 'dynamic balance' in what I call 'full cycle science'.

So, interestingly, Akihiro was connecting to the AR community in Australia at a time when it was ready for new life after the long tiring years of paradigm wars.

I mark the point of baton-passing from when we veterans may have 'peaked' in 1997 when thousands of us came together from numerous different fields of AR and PAR at Orlando Fals Borda's aptly named 'Convergencia' World Congress of PAR/ALARPM (Action Learning, Action Research and Process Management), in Cartagena, Colombia. And then from when we met for a time of historic fruition in 2000 at our Ballarat World Congress, attended by almost every major figure in global AR/PAR, and finally, at Jacques Boulet's 2010 World Congress of PAR in Melbourne, where some of us feeling we were now moving to being 'elders' amid a new generation who had not lived the battles we had lived, but were now facing new ones of their own resulting from ten years of world economic upheaval.

Perhaps if AR is always working to resolve conflictual differences and build life-giving change for our species by peaceful co-inquiry rather than warring pitched battles, it will always be both a hard road at the same time as incredibly exhilarating to be travelling companions with people on their-and-our voyages of life discovery. Akihiro has described one of my war stories, but, on reflection, perhaps we are in good company when we struggle, given that these were the same institutional forces that regularly array themselves against co-inquiry in the interests of power-exerting authorities, experts, leaders, and professionals – forces that closed iconic and successful centers, such as the Birmingham Cultural Studies Centre, responsible for the worldwide revival and dissemination of the discourse of 'culture', or our own incredibly influential and contributory Centre for Action Research and Professional Practise at Bath University in the UK, that had generated the monumental Sage Handbooks of AR and the journal of AR; and the similarly exceptionally valuable Cornell international PAR Network that had brought the voices of the colonized Latin America 'south' into a new participatory anthropology with 'the north'. Akihiro tells his students, 'we are not doing the wrong thing'. Indeed it seems very much the right thing if we want more 'fully living systems' While Akihiro points out the frequent impossibility of conversing with epistemic opposites, we can have some confidence that dialogue as exchange remains an underlying property of all living systems, so there will always be 'growing points' that seek to return to talking to dissolve the endless succession of force fields our species seems drawn to form around so many confounding bifurcations.

I look forward to many more such action research accounts as those coming out of Akihiro[2] Ogawa's quality course at the University of Melbourne that follow in this special *ALARj* issue.

References

Brown, L 1988, *Group self-evaluation – Learning for improvement*, Department of Education, Victoria, Melbourne.

---

2 The dictaphone just auto-typed Akihiro as: 'back a hero'. It seems the machines are becoming autopoetic and we are in dialogue!?

Kemmis, S, & McTaggart, R 1988, *The Action Research Planner* (3rd edn), Deakin University Press, Geelong.

Wadsworth, Y 2008a, 'Systemic human relations in dynamic equilibrium', *Systemic Practice and Action Research*, vol. 21, pp. 15–34. https://doi.org/10.1007/s11213-007-9080-6

Wadsworth, Y 2008b, 'Is it safe to talk about systems again yet?'–Self organ-ising processes for complex living systems and the dynamics of human inquiry', *Systemic Practice and Action Research*, vol. 21, pp. 153–170. https://doi.org/10.1007/s11213-007-9084-2

Wadsworth, Y 2011a, *Do It Yourself Social Research: The best-selling guide to doing social research projects* (3rd edn), Routledge, London.

Wadsworth, Y 2011b, *Everyday Evaluation on the Run: The user-friendly introductory guide to effective evaluation* (3rd edn), Routledge, London.

Wadsworth, Y 2010, *Building in Research and Evaluation: Human inquiry for living systems*, Routledge & Action Research Issues Association, London

Wadsworth, Y 2014, 'World congresses of action research', in Coghlan, D & Brydon-Miller, M (eds) The Sage *encyclopedia of action research*, Sage Publications Ltd, Thousand Oaks, CA pp. 826–829. Or open access summary: https://www.alarassociation.org/events/world-congresses, accessed 8/02/2022

Wadsworth, Y 2015, 'Shared inquiry capabilities and differing inquiry references: navigating 'Full Cycle' iterations of action research', in Bradbury, H (ed) *The SAGE handbook of Action Research* (3rd edn), Sage Publications Ltd, London, pp. 750–759.

Biography

Yoland Wadsworth (BA, PhD, FAES) is Adjunct Professor at Social & Global Studies Centre, RMIT University. She is the author of numerous books resulting from more than 20 years applied practice outside academe and more than 20 years within academe, as described in this autobiographical article. Some of these writings are indicated in the references above. See also https://livingsystemsresearch.com/ for a description of her magnum opus. Yoland is a former president of ALARPM, and has been awarded life membership of ALARA.

Research reflexivity: A journey of unlearning and co-learning

Ai Ming Chow

Abstract

Indigenous art markets in Australia are highly dynamic. Current literature on Australian Indigenous art markets reveals that they comprise of different stakeholder groups, including Indigenous artists, Indigenous communities, art intermediaries, community art centres, consumers, and art collectors. Upon reflecting on the settler colonial history in Australia that consists of continuous erasure of Indigenous people and their ways of life, it became important that I adopt a research trajectory that is participative, collaborative, and culturally sensitive. In this paper, I justify the importance of conducting participative action research which shares the spirit of the decolonization movements. Particularly, these movements advocate for an unlearning process that involves relinquishing of power. Then, I outline my research process that involves co-learning from different stakeholder groups through informal and formal field work. Lastly, I provide my reflexive account of the co-learning process as well as my positionality as a non-Indigenous researcher conducting research with Indigenous people and cultures.

Key words: Decolonizing research, Indigenous communities, Indigenous art market, relationality

> **What is known about the topic?**
>
> Conducting action-oriented research in the Indigenous communities and indigenous methodologies
>
> **What does this paper add?**
>
> This paper offers a reflexive account of research process that involves unlearning and co-learning within a relational indigenous paradigm
>
> **Who will benefit from its content?**
>
> - Graduate students of AR
> - AR dissertation writers
> - AR researchers engaging with Indigenous communities
>
> **What is the relevance to AL and AR scholars and practitioners?**
>
> - This paper offers the reflection and perspective of a graduate student's attempt and approach to conducting AR doctoral project.
> - The paper brings ALAR to the work of Indigenous arts in Australia.
> - This paper would be beneficial to AR dissertation writers about conducting action-oriented doctoral projects, particularly when they are engaging with Indigenous communities.

Received   January 2022   Reviewed July 2022   Published November 2022

Shifting from problem-focused perspective to relational paradigm

Aboriginal and Torres Strait Islander people began to occupy Australia more than 50,000 years ago. At the time of European colonization, an estimated 320,000 Indigenous people occupied Australia with the majority living in the country's south-east, and the Murray-Darling Basin (Australian Bureau of Statistics 2006). Colonization severely impacted Aboriginal society and economy. For example, there was epidemic disease that caused immediate loss of life, and the occupation of land by invaders and the restriction of Indigenous people to 'reserves' affected their ability to support themselves (Australian Institute of Health and Welfare 2015). The impact of colonization, relocation of Indigenous people to missions and the forced removal of children resulted in the dispossession and dislocation of Aboriginal and Torres Strait Islander people from their land and family. As a result, this further disrupted cultural beliefs and practices, and has adversely affected the social and emotional wellbeing of many Indigenous people (Swan and Raphael 1997). Until today, many Indigenous people continue to suffer from the consequences of European settlement, as some continue to live in conditions of economic disadvantage and marginalization. Acknowledging some of these issues in which large groups of Indigenous people in Australia are still grappling with, I decided to channel my time and resources as a doctoral student to understand the complexities inherent within these issues.

As I delved further into the history of colonization and its impact on Indigenous people and their cultures, I became fascinated with the development and dynamism of the Indigenous art markets in Australia. In 2020, Indigenous art continues to garner international interest with prominent exhibitions held at Gagosian Galleries in New York, Los Angeles and Hong Kong generating AUD $4.2 million of sales from 22 pieces of artworks (Coslovich 2020). Since European colonization of Australia, the meanings of Indigenous art have evolved. From being treated as ethnographic artefacts of 'dying' cultures in the early 19th century (McGregor 1997; Morphy 1998) to becoming valued commodities traded within the market (McCulloch 1999; McCulloch & McCulloch Childs 2008), different groups of stakeholders have played different roles and continue to shape the emergence of the market for Indigenous art. These important stakeholders encompass the Indigenous artists, commercial art intermediaries (i.e., galleries and private art dealers), public institutions such as museums, community-controlled arts centres, and governments (Wright 1999; Wright 2000).

Within the Indigenous art market, commercial art intermediaries are an important nexus that sit in a relational network, engaging directly but in different ways with Indigenous artists, their communities, community-controlled arts centres, and consumers. Since the 1970s, the Indigenous art market has evolved into an interdependent network of stakeholders who contribute to the production, distribution, promotion, and sale of the art of Australian Indigenous Peoples (Becker 1982; Boyd, Ward & Wright

2002). This art market increasingly operates as an 'art machine', largely driven by the rise of community art centres supported by government funding (Rodner and Thomson 2013). The metaphor of the art sector as a machine draws attention to the many interlocking stakeholder cogs, which include artists, curators, community art centres, art dealers, businesses, galleries, and consumers. However, in 2014, Woodhead and Acker (2014) estimated that less than 10% of people operating Indigenous art businesses are Indigenous people. While this does not diminish the achievements, resilience, and inventiveness of the Indigenous artists in the wider market, it shows that non-Indigenous stakeholders continue to play an important intermediary role within the market.

It is important to recognize and acknowledge these relationships among the stakeholders and communities, because, according to Chilisa, (2019), relationality is central across indigenous worldviews. While I continue to explore the general histories of Indigenous art markets in Australia and how it is shaped by legacies of colonization, I narrowed my focus on commercial art businesses. I became intrigued by how they position themselves within the Indigenous art markets, as well as how they engage with the Indigenous artists, communities and consumers. With such insights, I aspire to gain further knowledge about how we can transform the market space so that the commercial businesses are not dominated by non-Indigenous people but are driven by the agenda of the Indigenous people. I shifted away from studying the problems faced by Indigenous people which inadvertently would have framed the Indigenous people as the 'victims' of the situation that need 'saving'. Instead, I focused on seeking to understand the relationality among the individuals involved as we reimagine a marketplace where Indigenous people and their cultures continue to thrive.

Chilisa (2019) further highlights that, within a relational indigenous paradigm, research practices and methodologies should be guided by the histories, experiences, and cultural beliefs of Indigenous people. Therefore, it was imperative that I develop my cultural sensitivity towards the history and the cultures of Indigenous people in Australia. In this paper, I will highlight the key themes in my reflexive journey of conducting research with Indigenous communities and art intermediaries. In the following sections, I first broadly describe the decolonizing research methodologies, drawing parallels with action research. I will then describe how I build trust and connection with the selected groups of art intermediaries through *unlearning* and *co-learning*. Then, I will elaborate on my positionality as a non-Indigenous researcher conducting research related to the affairs and wellbeing of Indigenous people.

Research as a process of unlearning

Given that this study concerns the matters of Indigenous people living in a colonial settler society, I draw on decolonising methodologies (Denzin et al, 2017; Tuhiwai Smith 2012) to inform my choice of research techniques that are attuned to Indigenous cultural practices and perspectives when engaging with Indigenous people (Chilisa 2019). I have found these methodologies striking similar chords with the heart of action research, in which research aims to bring about positive transformation through engaging with stakeholders in the research process (Greenwood and Levin 1998; Ogawa 2006). Kovach (2021) further supports that community-based approaches are in alliance with the ethical and community dynamics of research with Indigenous people.

Decolonizing methodology

Decolonizing research requires acknowledgment of Indigenous ways of thinking, to build research knowledge that is culturally appropriate, respectful, honouring, and careful of the Indigenous communities and their cultures (Datta 2018). A sensitivity towards indigenous cultural systems is crucial because, despite well-intentioned interests, accomplishing altruistic goal for change is neither straight-forward nor apolitical. When conducting research within the context of Indigenous people, extra care is required to avoid privileging Western conceptualisations or ideas that could end up disregarding the interests or wellbeing of Indigenous people. For example, Tuhiwai Smith (2012) critiqued how knowledge about Indigenous people is collected and created, suggesting that it is often the case that knowledge extracted from Indigenous people may hurt Indigenous people, instead of improving their living conditions or well-being. As Tuhiwai Smith (2012) dramatically states,

> From the vantage point of the colonized, a position from which I write, and choose to privilege, the term *research* is inextricably linked to European imperialism and colonialism. The word itself, *research*, is probably one of the dirtiest words in the indigenous world's vocabulary (p. 1).

Researchers are often complicit in the removal and commodification of knowledge about Indigenous people (Denzin et al, 2017; Tuhiwai Smith 2012). Referring to research conducted within Indigenous contexts, Tuhiwai Smith (2012) makes a powerful case about how research is not an innocent academic exercise. Instead, she argues that we must understand the power relationships between the researcher and the researched. For example, when research methods used do not respect indigenous ways of knowing, this gives power to researchers who may misrepresent and stereotype Indigenous people, often denying them a voice in the production of Indigenous knowledge (Tuhiwai Smith 2012).

In the case of the Māori, for example, they have a naturally occurring way of sharing knowledge through oral narratives, or testimonies, that have an internal logic and rules; thus, Tuhiwai Smith (2012) encourages the use of more naturally occurring ways of collecting data that are culturally sensitive. For example, I used naturally occurring secondary data, such as Indigenous artists' and art intermediaries' media interviews and websites, to preserve their voices and stories. Instead of using research as a colonizing tool, decolonizing research methodologies focus one's effort to re-frame research as a site where

> …Indigenous and non-Indigenous scholars interact, share experiences, take risks, explore alternative modes of

interpretation and participate in a shared agenda, coming together in a spirit of hope, love and shared community (Denzin et al. 2008, 12).

This spirit is also emulated by action researchers who strive to engage with their stakeholders to bring about transformation and positive change. As a result, I decided to employ a multi-method research approach that captures different types of data sources, ensuring that I looked beyond any single perspective through exploring different and multiple data sources. Importantly, this research project (HREC Project Number: 1851203.1) has been approved by the Human Research Ethics Committee of The University of Melbourne whereby, details of the research process, confidentiality and data use, interview questions and possible effects and minimal risks were reviewed and approved.

Entering the community and building trusts

Through purposeful sampling, I dedicated extra attention and care in ensuring that I sought out both Indigenous and non-Indigenous gallery owners, while amplifying the voices of Indigenous people in different data sources, such as collecting Indigenous artists' writing in their websites, interviews, and press releases. I also followed Indigenous artists and activists on various social media platforms (such as Instagram), where they express their views, stories, and voices. Having both Indigenous and non-Indigenous participants involved in the research is important because decolonization requires shared commitment by both the colonized and the colonizers. There is a need to capture how non-Indigenous people working within the colonial structures think and act towards Indigenous people, as well as how Indigenous people view themselves in the world.

As a non-Indigenous researcher, I also actively reflected on my positionality in the research and my engagement with Indigenous research participants during my fieldwork and interviews by keeping a personal journal. I also organized informal dialogues with selected Indigenous scholars and activists throughout my data collection process to avoid privileging Western academic assumptions about research. Particularly, I organized meetings with Indigenous scholars as we reviewed my sampling procedures and interview protocols. These Indigenous scholars also suggested relevant reading materials that served as exemplars of research on Indigenous peoples and cultures to expand my perspectives of Indigenous knowledge systems. I also had informal dialogues with Indigenous activists who shared their experiences of being *researched on* by non-Indigenous researchers. These dialogues were helpful because they provided insights on how to conduct research that acknowledges and respects Indigenous cultures and ways of life. I acknowledge the significance of these conversations and relationships with these Indigenous scholars who have offered their wisdom to guide the inquiry of this study.

Datta (2018, p. 2) argues that "decolonization is an on-going process of becoming, unlearning and relearning regarding who we are as a researcher and educator and taking responsibilities for participants". Essentially, decolonizing research calls for a research agenda that is "sympathetic, respectful, and ethical from an Indigenous perspective" (Louis 2007, p. 49), privileging a "two-eyed seeing framework" (Bartlett et al. 2012, p. 335) of the Indigenous and non-Indigenous worldviews. It is also important to acknowledge that decolonizing research does not mean a rejection of all colonial methods and theories; rather it involves exploring a reconciliation of colonial and Indigenous knowledge. On a more practical level, it requires researchers to be reflexive of our traditional research processes and be open to making appropriate changes when Western methods are inappropriate or disrespectful towards Indigenous knowledge systems. In the following sections, I further elaborate on my research process and my research positionality.

Co-learning process through fieldwork

As part of my research journey, I carried out both informal and formal field work during the period of February 2017 and October 2019. Overall, I attended 50 different seminars, gallery openings, talks, presentations, panel discussions, and conferences. These informal and formal field works were central to facilitating the continuous engagement with the different groups of stakeholders and communities. Through the intimate dialogues and conversations with Indigenous artists, Indigenous Elders and their communities, scholars, and art intermediaries, they had shaped the trajectory of the research process. After every conversation and field visit, I kept detailed field notes which I shared with my co-researchers. While I am the graduate student who conducted the field work, my supervisors, who are also my mentors and co-researchers, continue to read my field notes as we meet regularly to discuss and reflect on our research process. For example, we challenged our views about what does Indigenous art mean, and we questioned our preconceived ideas of the value of Indigenous art. We also reflected on what would be considered 'best practices' in engaging with Indigenous artists and community members. The research process was emergent, allowing the underlying assumptions of the research methods to be critically examined and challenged iteratively. Overall, the research process essentially became a journey about *listening* to the priorities of the members of the communities (i.e., artists, intermediaries, and Indigenous Elders etc.). The focus was placed on what are the 'blind spots' that need to be addressed and changed within the existing market structures. It was less about me imposing my research agenda and the problems *I* wanted to solve.

Informal fieldwork

The purpose of the informal field work has been instrumental in educating me as I immerse myself in Indigenous cultures and in engaging with communities of Indigenous peoples, artists, scholars (Indigenous and non-Indigenous), and art intermediaries. The knowledge I have gained further enriched the dialogue I had with my supervisors who are also my co-researchers. In many cases, I became the 'teacher' while my supervisors became the 'student' because I was the researcher who had been immersing myself in my relationships and engagement with the communities. In February 2017, I was enrolled in a field work subject called 'Indigenous Perspectives on Development'. As part of a group of student researchers, we visited a town in the Northern Territory of

Australia, called Alice Springs, with a population of around 27,000, a third of whom are Indigenous people mainly of the Arrernte language group. For one week we also lived in a community called Ntaria (also known as Hermannsburg).

The community in Ntaria was an interesting place for conversations and field work because of its complex history of contact between Indigenous people and non-Indigenous people. More importantly, it is also the birthplace of the famous Indigenous artist, Albert Namatjira. Therefore, through this field trip, I revisited one of the key places where the art market emerged and where the Hermannsburg watercolour movement began. During this field work, I also visited the local institutions including the local schools, aged care centre, youth centre, as well as the famous Hermannsburg Potters. I had in-situ conversations with the artists about their art practices, stories and cultures. The pottery made by these famous potters were being sold at prominent galleries across Australia. We exchanged stories about life in Melbourne and life in Ntaria.

In addition to conducting field work at an Indigenous community in a remote area, I also attended seminars, lectures, panel discussions and talks by Indigenous artists in the urban area of Melbourne. At these events, I took detailed field notes to capture multiple perspectives, from Indigenous scholars, art historians, to Indigenous artists. These events provide rich social context to understand the different ways meanings of Indigenous art are constructed by various stakeholder groups involved in the market and how these meanings evolve over the years. These often become the focus of the discussions and meetings with my supervisors as co-researchers as we discover the heterogeneity in the meaning and value of Indigenous art.

This informal field work was also an important opportunity to network and identify the key galleries and prominent actors within the field of Indigenous art markets in Australia. For example, I attended 12 weekly seminars of a university subject called, 'Contemporary Aboriginal Art'. Through attending these seminars, I met an art curator, who then became one of the interviewees. In addition, as part of the subject, I attended an interactive seminar with an Indigenous artist who is a curator, Indigenous Elder and activist. During this seminar, he highlighted the differences between Indigenous knowledge systems and Western systems. He explained how these gaps and tensions had impacted some cultural practices and wellbeing of people in his communities. Subsequently, at other gallery openings, I met the two non-Indigenous curators whom this Indigenous artist has been collaborating closely with. These two art intermediaries also became interviewees for this study. Given that the Indigenous art world is a relatively small community, attending these events was crucial for me to engage with the relevant market actors, including art curators, gallery owners, and artists. I was also able to learn about important issues and the trends in the Indigenous art market, such as the prevailing issues with copyrights that affect the livelihood of many Indigenous artists. These interactions and relationships helped me and my co-researchers to expand on our perspectives and knowledge of the Indigenous art world.

Formal fieldwork

I also conducted formal field work at specific Indigenous art galleries and openings of new exhibitions across Victoria, Australia. I was a participant observer at these events and also conducted informal interviews in-situ. The focus of this data collection was to understand the art intermediaries' practices in engagement with different groups of market stakeholders including the artists and the consumers, trends, movements, and shifts within the Indigenous art market and to observe key stakeholders within these sites of commerce, art, and culture. Through the newsletters I have subscribed to as part of my archival data collection, I received frequent updates from galleries and art magazines (e.g., *Art Almanac*) about openings of new and exciting exhibitions.

At the exhibition openings, art curators and artists often provided more in-depth explanations of the artworks in the exhibitions. These openings generally attracted around 20 to 50 people and lasted about 2 hours. The majority of attendees are art collectors and potential buyers who are interested in the artists' work. The artists are often present to interact with the buyers and answer any questions. During these exhibition openings, I took detailed field notes on how the artworks were presented by the art curators to unpack the positionality of the art intermediaries as well as their relationships with the artists. I also collected printed and online exhibition catalogues of the exhibitions I attended. Most of the exhibition catalogues are publicly available online. These catalogues explain artists' works and art practices. Some of them also include essays by art curators and/or the gallery directors.

These openings and public events are social spaces where collectors, artists, curators, and gallery owners interact. Visitors must opt into the gallery's mailing list and often the people at the opening know one another. The people attending these openings were welcoming and many were incredibly passionate and wanted to share their love of art. Therefore, I generally found it easy to strike up a conversation about the artworks and their experience in collecting or buying Indigenous art. Also, given that artwork was on display, the art became the perfect conversation starter to engage with the collectors. I often prompted the conversations with simple questions such as, 'What do you think about this piece?', 'Is there any particular piece that you are interested in?' and 'What do you think about the artists' works in the exhibition?' These interactions were recorded in detailed fieldnotes immediately after each opening and public event. As a result, I was able to become more familiar with emerging and topical issues considered important among the key market actors. I was exposed to the important and emerging artists and art intermediaries within the local Indigenous art world. Moreover, I was invited to additional networking events, such as Indigenous artist's documentary screenings and talks. Through this process, I was constantly being challenged to confront my own biases, blind spots and my own positionality as a non-Indigenous researcher in the field. In the next section, I further elaborate on how I established my position as a non-Indigenous researcher in the research process.

Finding my researcher positionality

I am not a person of Indigenous descent. I was born and grew up in Malaysia, a country that is multi-ethnic and multi-cultural. However, Malaysia and Australia were both once controlled by British colonizers between the 18th and the 20th centuries. Unlike the settler colony of Australia, in Malaysia the land was returned to the original occupants of the country when the Federation of Malaya achieved independence on the 31st August 1957. During the British colonial administration in the 18th and 19th centuries, my ancestors migrated from China to Malaysia to work in tin mining factories and rubber plantations, in support of the colonizers' aim to extract resources from the land. Therefore, in terms of my ethnicity, I am a third-generation Malaysian Chinese who grew up in the postcolonial society of Malaysia. Despite being a person of Chinese descent, I grew up participating in native Malaysian culture. For example, *Bahasa Malaysia* (the language of Malaysia) is the national language of Malaysia. It is widely used in professional, legal, educational, and commercial contexts. In school, I became proficient in Bahasa Malaysia, English, and Chinese. I grew up learning about the native Malay culture while celebrating my own Chinese traditional customs. Growing up in a postcolonial society like Malaysia, my positionality may help me stay open-minded because I have lived through the decolonization process of the traditional owners of the land reclaiming their cultures and languages post-colonization.

I moved to Melbourne Australia for higher education in 2006. Throughout my education in Australia, I have learned little about the culture of Indigenous people. But when I studied the topic of smoking, I soon learned that high rates of alcoholism and tobacco consumption exist within some Indigenous communities. At the start of 2017, I participated in a field trip to Alice Springs, Northern Territory with academics and research students from the Faculty of Arts. The field trip at Hermannsburg opened my eyes to the various challenges faced by the Indigenous community. However, despite constraints placed on their livelihoods, Indigenous communities remained strong and resilient. Growing up in an independent postcolonial society (Malaysia) that celebrates and honours the cultures of different ethnicities, I became extremely curious about the evolving relationship between the Indigenous people and the non-Indigenous people who never left the country.

Reflecting on my own cultural background and ethnicity, I am neither an Indigenous nor non-Indigenous Australian. When I entered this field of study, I was an outsider. However, I approached this research with the desire to find the pockets of hope where Indigenous people are resiliently preserving their culture and actively maintaining their connection to the land. As Teila Watson, an Indigenous youth, courageously declared in a 2017 article for *The Guardian:*

> …our culture has always been resiliently dynamic in its ability to thrive, even among the depression of colonialism and attempted genocide. We may not now know all of our languages, songs, dances and stories, but many of us have had our knowledges translated and interpreted to us, and through us, our whole lives.

I am coming from a place of openness, with some naivety, but I seek to use my research to understand contemporary opportunities and tensions around Indigenous peoples' expression of cultures in the market, and by extension, the wider society. This intention focused my approach in collecting archival data, conducting field work and sampling of participants for this study.

However, given the limited exposure and understanding I had of Indigenous people and their cultures prior to conducting this research, I found myself struggling to reach out and connect with Indigenous people. It felt like there was a great divide between me (as a researcher) and them (as the researched). I realized that my view towards Indigenous people and their cultures was shallow and influenced by hearing about the persistent problems (such as addiction and alcoholism) in media news outlets and policy reports. While I was open to learning, I was problem focused on how I could offer solutions to these problems. Instead, I became aware that I needed to challenge and expand my perspectives of Indigenous people by educating and immersing myself in understanding the histories of the Indigenous people and their art. I also needed to remind myself about the resilience and strengths demonstrated by Indigenous people for millennia. In short, I needed to be more asset and capacity focused.

As a person who was unfamiliar with the art market and Indigenous cultures, I started by reading extensively on the history of the Indigenous art market in Australia. To be able to engage in meaningful conversations, interviews, and interactions with the key market actors, I needed to be aware of important historical moments and/or shifts in the history of the Indigenous art market in Australia. Also, I still needed some basic knowledge of the key artistic and stylistic movements that had emerged from different Indigenous communities across Australia. For example, informants casually referred to some significant Indigenous communities, such as the Yirrkala and Yolngu people from the Arnhem Land in the Northern Territory of Australia, or the key exhibitions that were game changers within the industry. This is where the archival data formed the foundation in which further insights were built upon and allowed me to delve deeper in conversations in the field and formal interviews.

Another important observation that I learned in the process is that these key informants were also curious about my positionality and stance towards the political and historical aspects of Indigenous art in Australia. Given that I am neither Australian-born nor a person of Indigenous descent, my ethnicity and cultural origins had positioned me in a peripheral space. I often felt like an outsider or an imposter in the room. I also had to manage the meaning of my younger age within openings of exhibitions full of more senior attendees. I believe that being in this position helped me enter the site with naivety and openness. I was not weighed down by some of the political and historical baggage often associated with non-indigenous Australians or in being perceived as the 'colonizers'.

Importantly, I had to constantly reflect on my positionality as a researcher trained within a Western educational institution, in which I had limited prior exposure to and experience in conducting research on Indigenous cultures and livelihoods.

Therefore, with the intention to *decolonize my mind*, I sought out support from Indigenous scholars at my department by organising meetings with them. I also sought guidance from non-Indigenous scholars who have gained trust and friendships in their engagement with Indigenous artists and their communities. I also kept a personal journal for reflections.

That being said, one of the key areas that required further improvement in my research process is the limited discussion on the perspectives of Indigenous people who have been involved in the project. While I had been consistently taking field notes of my experiences and sharing these with my co-researchers, we had not prompted the Indigenous people involved in the project to reflect on their involvement in the research. My conversations had predominantly focused on the topic of Indigenous art and marketing practices, I had not sought out further reflections on their involvement in this research. All of the Indigenous participants I engaged with had been generous in sharing their stories and perspectives in response to my interview questions. It is my intention to engage with them again after my doctoral study to seek out their perspectives on our research and how they have been impacted in the process.

Maintaining trust through unlearning, co-learning, reflexive of positionality

Overall, I took on a research strategy that emphasized the selection of data gathering methods that are aimed at maintaining trust, transparency, respect, and reciprocation between the researchers and the research participants (Tuhiwai Smith 2012). Greenwood and Levin (2007) further highlight that safe and trusting space needs to be created to facilitate the co-generative learning process between the researchers as outsiders and the local stakeholders as insiders. The data collection methods were shaped and informed by my knowledge gained through extensive and extended immersion within the communities of Indigenous artists, art intermediaries, and art experts. For example, through informal conversations, I started establishing connections with potential gate keepers beginning in February 2017 and continuing through to October 2018 when I did my first interview. These gatekeepers included art experts and art historians who passed on their wisdom and insights from working with the communities of artists and commercial art intermediaries. Some of their suggested practices were based on their personal experiences in the field, such as listening, staying open-minded, and acknowledging the diversity of Indigenous cultures. Similarly, these practices guided my interactions with the research participants. Importantly, these gatekeepers then introduced me to other art intermediaries, artists, and art experts in the field.

I also realized that it was important to develop a holistic understanding of the Indigenous art movement in Australia to show my respect and gain trust among the participants. Therefore, the archival data were purposeful in this regard. I was able to use this understanding to help build rapport with the participants and initiate deeper conversations about the role of art intermediaries. For example, from the archival data, I developed a better understanding of the range of artistic styles and painting materials commonly used by different communities of Indigenous peoples living in various regions of Australia. This knowledge helped me to probe art intermediaries on how they represent the diverse range of Indigenous artworks.

More importantly, this research approach also required active interrogation of my own standpoint, worldview, positionality, and research practices as a non-Indigenous researcher (Land 2015). In all my interactions with the participants, I shared openly about the research project and my journey of challenging my own biases and naivety. I approached every interaction with a beginner's mindset, making clear to the participants that I am not the expert, but they are. The interview became a co-learning process between me and my co-researchers, as well as with the interviewee as we exchanged stories about our experiences of navigating our positionalities and identities within the art worlds. This facilitated a shift in power dynamics in the researcher-and-researched relationship, which I found to be crucial to maintaining respect and trust among the participants.

Conclusion

The journey of conducting research that is grounded in co-generative knowledge and co-learning is often rife with complexities and richness. Greenwood and Levin (2007) provided one of the most memorable and humbling statements about carrying out action research:

> AR is not an ideal process, happening like neoclassical economics in an environment of perfect information, ceteris paribus, and other absurd non-existent conditions. It is a real process, happening in real-time contexts with real people and it has all the contingencies, defects and exhilarations of any human process. Dialectics may sound attractive, but often, as a lived experience, they are exhausting and even enervating. (p. 113).

Having reflected on my own research process, I can vouch that it is a journey well-worth the effort. Through my research process, it became clear to me how important it is to build trust and to learn together with the different groups of stakeholders involved. It is consistent across the action research literature that it is essential to involve all participants in Action Research activities (do Amaral & Okazaki, 2016; Greenwood & Levin, 2007; Thiollent, 2005). This calls for deep immersion with the relevant communities and continuous open dialogues. Particularly in studies that involve engaging with Indigenous communities, this requires the researcher to release one's hold of control and power over the research process. Instead, it opens up spaces where creative solutions can emerge from the people being affected by the persistent problems.

I have also learnt to be incredibly vigilant about my place in the research process, as a non-Indigenous outsider and as a researcher being trained in the Western institutions that possess the tendencies to perpetuate the legacies on colonization. Instead, through being reflexive and critical on my positionality and my research training, it turns the gaze on the constricting and colonizing structures within the research institutions and academy. In the spirit of being critical of the legacies of colonizing forces, I hope to open up alternative spaces of conversations and trajectories where Indigenous perspectives and knowledge systems are acknowledged and honoured. I seek to amplify the possibilities where Indigenous people continue to reassert their voices, power and wisdom within the wider societies.

References

Acker, T, 2016, 'Somewhere in the world: Aboriginal and Torres Strait Islander art and its place in the global art market', *CRC-REP Research Report CR017*, Ninti One Limited, Alice Springs, NT.

Arentos do Amaral, JA & Okazaki, E, 2016, 'University students' support to an NGO that helps children with cancer: Lessons learned in thirteen academic projects', *Apoyo de estudiantes universitarios a una ONG que ayuda a niños con cáncer: Lecciones aprendidas en trece proyectos académicos.*, International Journal of Action Research, Rainer Hampp Verlag, vol 12, no. 1, pp. 38-58. doi:10.1688/IJAR-2016-01-Arantes

Australian Bureau of Statistics, 2008, *Population characteristics, Aboriginal and Torres Strait Islander Peoples 2006*, ABS cat no 4713.0.

Bartlett, C, Marshall, M & Marshall, A, 2012, 'Two-eyed seeing and other lessons learned within a co-learning journey of bringing together Indigenous and mainstream knowledges and ways of knowing', *Journal of Environmental Studies and Sciences*, Vol. 2, no. 4, pp. 331–340.

Becker, H.S, 1982, *Art worlds*, University of California Press, Berkeley, California.

Boyd H, Ward, S & Wright, F, 2002, 'The Indigenous visual arts industry, in Altman, J & Ward, S (eds.) *Competition and Consumer Issues for Indigenous Australians*, Australian Competition and Consumer Commission, Canberra, pp. 64-101.

Bristor, J.M & Fischer, E, 1993, 'Feminist thought: Implications for consumer research', *Journal of Consumer Research*, vol. 19, no. 4, pp. 518–536.

Chilisa, B, 2019, *Indigenous Research Methodologies (2nd ed.)* Sage Publications Inc., Thousand Oaks, California

Coslovich, G, 2020,, 'Gagosian opens $5m Indigenous art exhibition in HK', Australian Financial Review, Sept 30, https://www.afr.com/life-

and-luxury/arts-and-culture/gagosian-opens-5m-indigenous-art-exhibition-in-hk-20200929-p560ef

Datta, R, 2018, 'Traditional storytelling: An effective Indigenous research methodology and its implications for environmental research', *AlterNative: An International Journal of Indigenous Peoples*, vol. 14, no. 1, pp. 35-44.

Denzin, NK, Lincoln, YS & Tuhiwai Smith, L, 2008, *Handbook of critical and Indigenous methodologies*. Sage Publications Inc, Thousand Oaks, California.

Dodson, M, 2003, 'The end in the beginning: Re(de)finding Aboriginality', in Grossman, M. (ed.) *Blacklines: Contemporary Critical Writing by Indigenous Australians*, Melbourne University Press, Melbourne, Australia, , pp. 25-42.

Greenwood, DJ & Levin, M, 2007, *Introduction to action research*, Sage Publications Ltd., California.

Kovach, M, 2021, *Indigenous methodologies: characteristics, conversations, and contexts* (2nd edn), University of Toronto Press.

Land, C, 2015, *Decolonizing solidarity: Dilemmas and directions for supporters of Indigenous struggles*, Zed Books Ltd., London.

Louis, RP, 2007, 'Can you hear us now? Voices from the margin: Using Indigenous methodologies in geographic research', *Geographical Research*, vol. 45, pp. 120- 139.

McCulloch, S, 1999, *Contemporary Aboriginal art: A guide to the rebirth of an ancient culture*, University of Hawai'i Press, Honolulu, HI.

McCulloch, S & McCulloch Childs, E, 2008, *McCulloch's contemporary Aboriginal art : The complete guide* (3rd edn), McCulloch & McCulloch Australian Art Books, VIC, Australia.

McGregor, R, 1993, 'The concept of primitivity in the early anthropological writings of A.P. Elkin', *Aboriginal History*, vol. 17, no. 2, , pp. 95–104.

Morphy, H, 1991, *Ancestral connections: Art and an Aboriginal system of knowledge*. University of Chicago Press, Chicago..

Ogawa, A, 2006, 'Initiating change: Doing action research in Japan', in Gardner, A & Hoffman, DM (eds.), *Dispatches from the Field: Neophyte Ethnographers in a Changing World*. Waveland Press, Long Grove, IL, pp. 207–221.

Rodner, V. L & Thomson, E, 2013, 'The art machine: Dynamics of a value generating mechanism for contemporary art', *Arts Marketing: An International Journal*, vol. 3, no. 1, pp. 58-72.

Raphael, B & Swan, P, 1997, 'The mental health of Aboriginal and Torres Strait Islander People', *International Journal of Mental Health*, vol. 26, no. 3, pp. 9-22. http://www.jstor.org/stable/41344833

Thiollent, M, 2005, 'Insertion of action-research in the context of continued university education', *International Journal of Action Research*, vol. 1, no.1, pp. 87-98.

Tuhiwai Smith, L, 2012, *Decolonizing methodologies: Research and Indigenous Peoples*, Zed Books, London.

Woodhead, A & Acker, T, 2014, 'The art economies value chain reports: Synthesis', *CRC-REP Research Report CR004*, Ninti One Limited ,Alice Springs..

Wright, F, 1999, *The art and craft centre story volume one: A survey of thirty-nine Aboriginal community art and craft centres in remote Australia*, Aboriginal and Torres Strait Islander Commission, Canberra:

Wright, F, 2000, *The art and craft centre story volume three: Good stories from out bush – Examples of best practice from Aboriginal art and craft centres in remote Australia*, Aboriginal and Torres Strait Islander Commission, Canberra:

Biography

Ai Ming Chow studied the Indigenous art markets in Australia, focusing on commercial art intermediaries. She specializes in alternative methodologies such as interpretive, critical, participatory and community action research methods. Her scholarship has appeared in academic journals such as Journal of Customer Behaviour, in conference proceedings for Association of Consumer Research (ACR) Conference and Consumer Culture Theory (CCT) Conference. She is currently working as a Melbourne Early Career Academic Fellow (MECAF) at the Department of Management and Marketing (Faculty of Business and Economics), The University of Melbourne.

'She might find out the truth:' Action researching with young theatre audiences

Abbie Trott

Abstract

Action Research is often used in theatre and performance to engage communities in acts of social change. Framed as theatre action research in applied theatre, practice as research or participatory enquiry, theatre becomes the central tool in generating knowledge. However, engaging with audiences experiences of theatre as an act of performance throughout my PhD means I was interested in how AR's paradigmatic qualities apply to young people's experience of theatre. Densely detailing my methodological approach, in this paper I examine how AR applied across my research and has subsequently become embedded in my methodological approach to analysing theatre and performance.

Key words: Theatre, action research PhD, theatre action research, knowledge generation

What is known about the topic?

Reason (2006) talks about using action research with young audiences

What does this paper add?

This paper takes the reader through the experiences of an action research focused Theatre Studies PhD

Who will benefit from its content?
- Theatre studies scholars
- PhD candidates and other students
- Action Researchers interested in theatre and performance

What is the relevance to AL and AR scholars and practitioners?
- The paper is relevant because it offers emerging AL and AR scholars and dissertation writers a view into the nuts and bolts of an Action Research focused PhD
- It actively links Action Research to theatre studies and performance, beyond applied theatre.

Received January 2022   Reviewed June 2022   Published November 2022

Introduction

'She might find out the truth' was part of a young person's description of a photo representing their memories of what *Hungry Ghosts* by Jean Tong was about. The play was performed as part of the Melbourne Theatre Company's education program for 2018 and traced the journey of the character '2' as she traversed a new life in Melbourne, the missing MH370 aeroplane and Malaysian governmental corruption. The image this young person chose to caption (see figure 1) is of '2' at the end of her journey and arguably reflects the point at which she discovers herself; discovers if she was, or wasn't on that fateful plane, and the nature of her disconnection from her home country of Malaysia—as the second part of the caption notes 'about the MH370 missing.' The meaning this young person was able to resurrect about the play, three months after they saw it, is poignant. It demonstrates that with careful drawing out of memory they could link together multiple strands of the performance and make sense of the world. It is just one example of how the young people I researched with over my PhD created their own meaning.

Figure 1 Emina Ashman playing '1', Bernard Sam playing '3' and Jing-Xuan Chan playing '2' in Hungry Ghosts by Jean Tong. © Melbourne Theatre Company. Photo by Jeff Busby (2018)

As research participants, young people are at the heart of my project. My PhD began by questioning how young people — teenagers specifically — engage with theatre in the age of ubiquitous media: We are surrounded by the digital, and young people (aged 14-19) have come of age with smartphones in hand. Referred to by a variety of monikers; Gen Z (Dorsey 2016, p. 14), 'iGen' (Twenge 2017, p. 8), 'Net Generation' (Tappscot quoted. in Lewis & Johnson 2017, p. 124), or 'iGeneration' (Rosen quoted. in Lewis & Johnson 2017, p. 124), the relationship between young people and the digital culture is complicated. Regardless, I think it is fair to say that young people know how to negotiate the digital culture. This prompted me to question if these strategies of negotiation were translatable to the experiences of young theatre audiences. As audiences of theatre, young people are doubly disadvantaged when it comes to the making of meaning: they are young, and they are audiences. Hence, it became apparent to me early on that if young people were going to be at the centre of my research, they needed to be afforded the agency to fully participate. Affording them the agency to participate raised a series of methodological challenges, which I outline in the paper.

As a project reliant on the authority, agency, and sense of belonging in the participants it became apparent to me that my PhD research paradigm and methodology needed to focus specifically on these factors. A passing mention of action research (AR) in an article by Mathew Reason (2006) introduced me to how AR can be used in a theatre studies project. Drawing methodologically on the tenets of AR, Reason and his team sought to research 'with,' rather than 'on' young people about how live theatre impacted on them (2006, p. 132). Running alongside was the research tools my colleagues were developing through the Creative Convergence ARC Linkage research project[1] my PhD project was positioned within. Rachel Fensham and Megan Upton describe this process as one where young people were given "permission to assess objectively and collectively the role and purpose of their memories as an experience of theatre" (2022, p. 7). In undertaking a project that investigated *with* young theatre audiences, I was drawn to the paradigm and joined Akihiro Ogawa's postgraduate course work subject in 2017. There I became acquainted with the knowledge generation, living systems and cyclic research that hold an AR approach together. Across my project, these three tenants of AR were useful in guiding the ways that my methodology developed, and the analysis that I could undertake. By applying AR ideas around knowledge generation, living systems and cyclic research, my research could be guided in

---

1   An ARC (Australian Research Council) linkage research project is a federally funded research project with connections to industry. Creative Convergence (2017-2022) had 10 industry partners, all focused on theatre and young people from regional Victoria.

such a way that the young people at the heart of the research were afforded agency and participatory authority.

Researching with my participants, three central questions emerged: How do I engage with young people in a meaningful and respectful way? How do I ensure that theatre stays the subject, not the tool of my study? And, how do I ensure that the reflexivity of my research carries through? To address these questions, across my PhD project I methodologically applied participatory modes of engagement influenced by an AR paradigm. In actively engaging with how knowledge is generated, how theatre research is a living system, and the centrality of a cyclic approach to empowering young people to act with agency I was able to answer these questions. I applied the tenets of AR to four case studies, and the methodological approach I developed grew and I refined the knowledge I was building. In this paper, I investigate the philosophical underpinnings of AR across the length of my PhD project where I examined the reception of cultural artefacts, specifically the theatre productions that were at the heart of my PhD. I use the three questions as a guiding framework off which I hang detailed descriptions of my process.

The four performance case studies at the heart of my project all engaged with young people in different ways. *Hungry Ghosts* (2018) was a theatre production directed at students in the last two years of high school. Also aimed at high school students, *Take Over: A Digital World* by Geelong Arts Centre (2018) engaged young people in the making and presentation of a theatre festival. In contrast *As If No-One Is Watching* by Vulcana Circus (2018, 2019) was a multi-generational women's circus performance and *Body of Knowledge* by Samara Hersch (2019) was developed with teenagers but for an adult audience.

Methodological considerations: Engaging with young people in meaningful and respectful ways

Outside the different way my case studies engaged with young people, there are several considerations which impacted on the methodological choices and context that surround my research. Firstly, the changing face of audience reception studies over the last decade privileges the audience's role in the making of performance. Secondly, the importance of recognising the agency and empowerment of audience members, especially young audience members, in that theatrical experience. Finally, the need to approach what Kirsty Sedgman describes as the 'multifarious' nature of theatre and performance reception research with rigour (2019, p. 463). It, therefore, became apparent that the research project needed to draw on multiple voices which would provide rich, thick data to understand and respond to Sedgman's question, '[d]o we know *how* theatre matters to them?' (2016, p. 90; emphasis added).

Broadly, knowledge is experienced closely with power and agency. The more knowledge one holds, the more authority they wield. Robin Nelson opens his discussion concerning practice-as-research in theatre with Jean-François Lyotard's musing; 'knowledge and power are simply two sides of the same question' (2006, p. 105). While Nelson is discussing the generation of knowledge in practice-as-research, the claim also applies in theatre more broadly. Knowledge holders in theatre have historically been considered the writers and directors, and more recently, the performers. As a body in the performance space, the knowledge the 'audience' holds has historically been perceived as having less worth. Within theatre studies, historically, the audience was subordinate to the other stakeholders (Freshwater 2009, p. 7). With changes in audience reception studies however, the knowledge an audience brings to performances they spectate now holds value, and knowledge is one factor that audience reception scholars commend as important to engagement.

While these elements all trace across the reception of theatre, they are also integral to researching reception, and as a result, these findings must be intertwined with their methodological choices. In her 2019 article, *On rigour in theatre audience research*, Sedgman proposes that '[f]indings – and interpretations of findings – should always be presented through the lens of methodology' (2019, p.

476) when examining how audience studies in theatre can find its way through the 'political, epistemological, ethical, and practical implications' inherited from the previous generations of how theatre scholars engaged with theatre audiences (2019, p. 464). Building on this, it follows that knowledge, belonging and understanding the terms of reference which are so integral to young people engaging with theatre must also be integrated into any methodological approach applied in a reception study.

Theatre and young audiences

Theatre audience reception studies shifted in 1990 when Susan Bennett used theories of spectatorship and literary reader response theory in her analysis of theatre audiences (Freshwater 2009, p. 12). Martin Barker writes that until Bennett's book 'there has been little attempt to see how actual, differentiated audiences respond' (2003, p. 20). It is this sense of privileging the experience of the individual which has shifted to the forefront of contemporary audience reception. For example, Australian theatre audience scholar Caroline Heim situated the empirical embodiment of the audience's role in what she describes as mainstream theatre (2017, p. 2), and one research methodology she uses is to engage with actors, ushers and other theatre professionals about their observations of the audiences' experience. As Heim states, 'the actors are one of the audience's observers' and this situates the audience as performer (2017, p. 3). Heim juxtaposes the voice of the actor, with their 'insights' and 'understandings' of the audience, with the voice of the usher, who, in her words lives 'in the liminal space between the stage and the audience' (2017, p. 8).

While the audience is central to this picture, as I alluded to at the beginning of the paper, the participants in this research were disadvantaged because they were audiences, and because they were young. Belonging, community and social inclusion are regularly found to be central to the ways that young people engage with theatre (Baxter et al. 2013; Brown & Novak 2007; Cultural Development Network 2016; Foreman-Wernet & Dervin 2013; Reason 2010). There are three key studies that have used a participatory approach with young audiences: Mathew Reason,

John Tulloch and Kathleen Gallagher. Reason and his team engage the young people with the reporting and knowledge generation by discussing questions of liveness that were posed by the researchers, and the research (Reason 2006, p. 140). Reflecting, that if a methodological goal of the project was to enhance young people's understanding of liveness, Reason argues that participatory research methodologies applied were effective because of the relationships formed between researchers and participants (2006, p. 143). Central to Tulloch's methodology was understanding how the students' knowledge affected their engagement with 'high culture' (Tulloch 2000, p. 85). Tulloch proposed that holding either 'expert (official) or lay (everyday) knowledge,' enhanced the effect of the performance, and their engagement with it (2000, p. 89). Undertaking a longitudinal pedagogical study of high school students in India, Taiwan, Canada and the United States, Gallagher's team offer key insights into methodological considerations to measure the engagement of young people in theatre studies and drama (Gallagher 2014). Developing a methodology that locates the student voice as central to engagement Gallagher's research provides another example of the importance of positioning young people as co-researchers and using a mixed methodology to engage and empower the young participants (2014, p. 11). Each of these studies approaches understanding, agency and belonging being afforded to young theatre audiences, as aspects of knowledge generation. The young people needed to be positioned as the expert of their own experience.

Co-generating knowledge

Taking these factors into consideration, my approach to using an AR paradigm is that knowledge is co-generated. Participants and researchers work together to generate the knowledge building on previous stages constantly evolving in what Yoland Wadsworth (2011) describes as an autopoetic living system. In asking young people to recollect, for example, the lingering memories of performances, I was asking them to generate a picture of the performance based on its affect and their memory. This became a living system, reliant on autopoiesis, or 'self-making,' because the co-generated knowledge was made up of individual 'sub-systems' (Wadsworth 2011, p. 41), with each sub-system integral to the development of the larger one.

Individual perspectives become integral to the system of the research, because as Davydd Greenwood and Moreton Levin frame it 'AR is based on the affirmation that all human beings have detailed, complex, and valuable knowledge about their lives, environments, and goals' (2007, p. 104). Greenwood and Levin (2007) argue that

> the crux of the AR knowledge generation process is the encounter between local insights and the understanding that the outsider brings to the table and the fusion of these insights into a shared understanding that serves as the basis for solving practical problems. These two forms of knowledge both connect and are quite distinct (p. 102).

When the system is 'living' then the individual contribution of each 'sub-system' becomes pivotal to the culminating 'body of knowledge' that can be assembled. If knowledge is at the centre of research, an AR approach to knowledge generation fields a cyclic reflexive process where each 'phase' of knowledge generation (research) is reflected on, and any learnings implemented in the next phase as an autopoetic system. In an AR paradigm, knowledge is often understood as belonging to insiders and outsiders.

Greenwood and Levin (2007) identify that there are two 'groups of actors' in a cogenerative model of research – outsiders and insiders. As the 'owners' of the problem, participants are the focus of the research as insiders. The professional researchers 'seek to facilitate a colearning process' as outsiders to the problem (Greenwood & Levin 2007, p. 93). Through a cogenerative learning process, outsider knowledge clarifies insider knowledge to create a mutual understanding of the 'problem' (Greenwood & Levin 2007, p. 94). Each group of 'actors' brings with them different knowledges: grounded research by insiders, or broader and more targeted in the act of research by outsiders. This creates what

Greenwood and Levin describe as an 'asymmetry in skills and knowledge' whereby it is the responsibility of the outsider to ensure that the insider gains the skills they need to cogenerate knowledge (2007, p. 95). This dynamic tension is the 'basis for [the] cogenerative process' (Greenwood & Levin 2007, p. 107).

In aligning AR with the theatre reception process, I investigate how a co-generating, autopoetic living system is translatable to how theatre audiences generate meaning through their relationship of insiders and outsiders to the works being performed on stage. A theatre performance is an act already aligned to being understood as knowledge co-generation. Each performance is live, and each live performance occurs with a different audience. As such, the meaning made in each performance can be nothing other than unique. Taking this to the extreme are interactive performances such as *Body of Knowledge*. In this performance, the audience is led through a series of acts and asked a series of intimate questions by the performers who are telephoning in from remote locations. In the end, the answers to the questions are reflected back to the audience, and that 'body of knowledge' built by the audience and their collaborating performers is unique.

Theatre action research

But what does this acknowledgement of co-generators knowledge mean for researching with audiences? While participatory enquiry is becoming common in audience reception studies, AR is rarely overtly applied as a research paradigm in theatre studies. When it is, it is usually within applied theatre, where theatre scholars and practitioners deploy theatre as a tool within projects – usually educational, community or therapeutic – whose participants do not necessarily see themselves as artistic or theatrical. Fitting within a broader participatory research model, emergent between AR and applied theatre, is what James Thompson (2012) has defined as Theatre Action Research (TAR). Similarly to AR, TAR is invested in democracy and change, and deployed to understand what is actually happening on the face of ethnographic and applied research where 'theatre itself is the research process ... *theatre* is an *action* that is *research'* (Thompson 2012, p. 122). Thompson and other researchers applying an AR framework in applied theatre such as Dwight Conquergood (2002) practice a process that involves workshops, content generation and performance of theatre, for, by, and with community members. Framed as 'radical research' Conquergood suggests a model of ethnographic research which breaks down the boundaries between the objective researcher and the 'subject' focusing on a way of 'knowing how' and 'knowing who' as opposed to 'knowing that' and 'knowing about' (2002, p. 146).

Through his work in refugee camps, where he employed theatre making workshops in a social change setting, Thompson (2012) applies theatre as a social research method and theatre itself is the research process. He argues that because theatre 'is where people's own stories can be presented, heard and transformed' (Thompson 2012, p. 122), it exists on the AR spectrum. This is where AR and applied theatre meet; positioned within a spiral of research, where each phase informs the next. For Thompson, 'TAR is thus the use of the body and speech to demonstrate and explain to 'critically examine'' (2012, p. 125). Using tools that are not just a 'means of intervening in community and group development' but 'an invaluable tool of participatory research' (Thompson 2012, p. 126), Thompson shifted the focus of theatre research.

Engaged with similar practical work to Thompson, Conquergood (1988) took a philosophical approach and theoretically applies the tenants of AR to how meaning is made in theatre. For example, he cites James Clifford when he argues that refugees in refugee camps exist in a liminal space: 'Betwixt and between worlds, suspended between past and future, they fall back on the performance of their traditions as an empowering way of securing continuity and some semblance of stability' (Clifford quoted in Conquergood 1988, p. 180. As such, the playfulness of creativity rooted in performance allows them to invent new cultures of engagement. For Conquergood, ethnographic fieldwork is an embodied act, positioning the body as central to the enquiry, and the drawing together of AR and Applied Theatre practice foregrounds the sensuous nature of embodied research, acknowledging 'the interdependence and reciprocal role-playing between knower and known' (1988, p. 182). Subsequently, liminal spaces — such as the refugee camps he worked in — where identity is a performance of boundary, and meaning is founded in relationships (Conquergood 1991, p. 185). In this he drew the work of Victor Turner and Nancy Fraser firmly into performance studies and applied theatre, situating TAR as a model of engagement which privileges the role of the knower in knowledge generation.

Using participatory data collection tools which Conquergood (1991) and Thompson (2012) situate as part of TAR, allowed me to ensure that the young people were empowered and able to act with agency. Taking the participants back into the liminal space of the theatre — like the liminal space of the refugee camp — allowed them to recall their embodied experience of watching and listening. Further, in allowing them to generate the lingering meaning of the performances they had watched, I prefaced their role as collaborators in the generation of knowledge. Speaking with the *Take Over* participants, together they generated a recalled picture of the performances they had seen. As a collective, they remembered in a workshop that the actors in another performance "put these helmets on" as they reproduced a moment of performers putting on clear safety glasses representing virtual reality (VR) headsets, a movement they replicated more than 12 months later when I visited them and asked them to recreate what they remembered (6 Aug 2018 and 10 Sept 2019). As audience members, their role in the meaning being made was central to the performance that they watched, and they became the outsider of the autopoietic system.

Through affording each audience member agency and authority, the experience of each individual — as part of a collective — can be prioritized and a participatory method ensures that knowledge is collectively generated. Reason and his team developed a methodological approach to investigate the impact of theatre on high school aged students that 'consciously and very deliberately recognized the participants as active audience members and individuals' (Reason 2006, p. 132). And 'active' is the key word here for how Reason argues his project is aligned with AR, and what I wanted to carry across into my research, this idea of the active participant. Reason (2006) argues that the transparency of his research motivations and focus on active participation ensures he is able to ethically apply an AR model, despite his imposition of questions onto participants (p. 132, footnotes 7). One method he uses to promote this is to get consent from his participants to use their correct names in the research, this means that they can see themselves on the page. Due to the ethical requirements of my institution, I was not able to ask participants to use their own names, but I did ask them to choose a pseudonym. While this compromise did shadow their contribution, they are still able to see themselves in the research.

Case study	Subject	Object	Data Collection
Hungry Ghosts	Young People	As audience	Workshop
Hungry Ghost	Theatre Makers	As makers	Observation / Interview
Hungry Ghosts	Self	As audience	Performance analysis
Take Over	Young People	As audience	Workshop
Take Over	Young People	As makers	Interview
Take Over	Self	As Audience	Performance analysis
As If No-One is Watching	Theatre Makers	As makers	Interview
As If No-One is Watching	Theatre Makers	As audience	Interview
As If No-One is Watching	Self	As audience	Performance analysis

Case study	Subject	Object	Data Collection
Body of Knowledge	Self	As audience	Interview / Performance analysis

Table 1. Data collection tools, subjects and objects.

My methodological approach

My research about AR, TAR and PAR enquiries with young theatre audiences highlighted the need, methodologically, to foster a sense of agency and embed a feeling of belonging, and while mixed, the approach I took was multimodal, and centred in the three tenets of action research I outlined earlier: knowledge generation, living systems and cyclic research. There were seven overlapping and iterative phases to my research: observation of creative development and rehearsals; pre-performance interviews with performers and directors; pre-performance workshops with young audience members; analysis of the performances; immediately post-performance vox-pop interviews with young audiences members; interviews with actors and directors in the weeks and months after the performances; and, post-performance workshops with young audience members between three weeks and 16 months post-performance. This research methodology drew on the one developed as part of the Creative Convergence research project my PhD sat within, and specifically examined the "performative role of memory" in theatre for young people (Fensham & Upton 2022, p. 2). I applied data collection tools outlined above to each of the case studies differently across the two and a half years of my fieldwork.

The first data collection point was the observation of *Hungry Ghosts* and *Take Over* rehearsals between 2017 and 2018. Next, my colleague Paul Rae and I interviewed the *Hungry Ghosts* cast before the performance season began in 2018, but I visited the *Take Over* participants on my own where I interviewed two school groups immediately before their performances. After reflecting on the initial stage of observation, I ran a series of workshops for *Hungry Ghosts* audiences. The first of these was a pre-performance workshop with young people associated with the theatre company. Immediately after we saw the opening night of *Hungry Ghosts* together, I ran a Vox Pop activity with the same group of young people. The next phase in my research consisted of a performance analysis of all four performances. When I watched *Take Over* and *Hungry Ghosts,* I took notes during the performances, and I had access to archival video recordings which assisted in my analysis. When I first saw *As If No-One is Watching* in 2018, I took very brief notes after the performance. The second time in 2019, I was more comprehensive in my note taking making sure I at least recorded who I saw and what they were doing. Occurring at the very end of my fieldwork, I applied a more systematic approach to my performance analysis of *Body of Knowledge* and developed a series of prompts that I responded to immediately after the performance. These prompts reflected the Vox Pop exercise I had run with the young people about their experience of *Hungry Ghosts*. In my observations of *Body of Knowledge*, I was most interested in assembling data that only related to the experience of watching; I wanted to replicate as accurately as possible the post-performance process I had deployed with *Hungry Ghosts* audiences to examine my own experience of being the audience, making me a research participant.

In addition to the Vox Pop activity, the post-performance data collection activities consisted of an open-ended interview with the *Hungry Ghosts* cast, and interviews with one *As If No-one is Watching* performer, who also watched the second season, and one director. I ran workshops with for *Hungry Ghosts* and *Take Over* audiences, each held between three weeks and sixteen months after they had audienced and performed. The first workshop participants were young people I had watched *Hungry Ghosts* with three weeks before. Additionally, four months after they saw the performance, Rae and I ran a research workshop with twenty-four high school students who had audienced *Hungry Ghosts*. The final workshop participants were with the two *Take Over* school groups I had interviewed before their performances in 2018. I visited both groups in 2018 and then once again in 2019 because I wanted to speak to the same young people more than once. The workshop model draws on the research methodology developed by Rachel Fensham, Paul Rae and Meg Upton into the impact of theatre on regional young people.

Theatre as the subject, not the tool of research

As a theatre and performance studies scholar, interested in theatre as the subject of the research, my approach was engaged with how action and participatory models in AR can be exercised in theatre reception studies. With a focus on audiences, I wanted to discern how performance matters to its audience where theatre is the object of the study – regardless of the tools adopted to understand its effect on audiences. In using a TAR approach there is the risk that it is centrally positioned as the research tool – a positioning contradictory to my placement of theatre as the subject of the research. Action and participatory models of research are used in theatre reception studies to discern how performance matters to its audience, so theatre is still the object of the study; regardless of the tools used to understand its effect and affect on audiences.

These reflections about how AR might be used in theatre studies highlighted the challenges I faced. My project concentrated on the experiences of young theatre audiences; however, the analysis of theatre as a cultural artefact needed to be the focus. I needed to find a way to draw on AR in both the ways that I engaged with participants and how I engaged with performance as the *subject*, not the *tool* of the research. Questioning the separation of theatre and tool speaks to the ongoing challenge that AR faces as it navigates the relationship between content and process but offers a different perspective. Rather than the subjectivity of the participant being questioned, rather the data collection strategies are doubled, and the object becomes the subject, and the content the process. Applying the principles of AR to my research with those young audiences ensured their participation through multiple stages of data collection. While the young people or theatre makers would not be able to direct the realisation of the research goals, their participation was conceivable as empowered and engaged. On reflection, I also realised that I was applying the outcomes of one phase of research, to the next, and so creating a three-cycle autopoietic system of knowledge generation to the PhD process.

In the workshop phase of the research involving *Hungry Ghosts* and *Take Over* I described earlier, I focused on situating the young people as the expert of their experience. Central to these workshops was an attempt to understand what about the performance affected them. Run at least weeks if not months after they saw the performance, I was interested in what they remembered, what stuck with them over time, what *affect* the performance had. Part of my process was to actively play down my knowledge of the production, to ensure that they felt comfortable in the act of remembering. For example, in one of my workshops where we were investigating their memory of a performance from a different group on the night they performed on, the following exchange occurred;

> Participant R: I don't know, I think she snapped out of it at the end.
>
> Participant N: How did it finish? Did they all come back out?
>
> Participant R: I think that she, that he went back to normal like regular self.
>
> Participant O: I think they could take their helmets off.
>
> Participant R: Yeah
>
> Participant O: Or didn't they do the start again? Like 'one player connected' [computerised voice]
>
> Participant N: Did they?
>
> Participant O: I feel like they did the start …
>
> Participant R: Meh
>
> Participant O: … the start again
>
> Participant N: I can't remember
>
> Participant R: It was a while ago.

Abbie:	It was a while ago, and did they do their tech rehearsal the same day that you did?
Participant N:	Yeah, that's why we remember it so well. But it's confusing because we are remembering two versions of it.
Abbie:	We don't know how it ends.
Participant N:	Do you remember how it ends?
Abbie:	No, I forget too.

[general laughter].

During these sessions, I would draw their recollections and re-creation out by asking them 'what happened next?' 'What do you remember next? 'and then 'what did they do?' 'and then?' 'and next?' I would at times recap what they had asked as a further strategy to draw their recollections of the performance. An example of this is the following:

Participant N:	Like she's trapped.
Participant O:	It was like we were trapped as well, like we couldn't get out of the game.
Abbie:	So, she worked out she was trapped and then what happened?
Participant R:	They all went into the first game?

I was very conscious of trying to ensure that they were recollecting the performance they had seen, without leading with my recollections. This strategy was to ensure that they realised that I was not the expert either; we were in this together. In speaking directly about my lack of expertise, I was able to reassure the participants that their contributions were valuable. I was overt in letting them know that I was not the expert, and I was not interested in my recollections. Speaking to a lack of expertise was challenging as an educator and scholar, especially within a classroom setting where students are conditioned to accept their educators as the expert. However, empowering the young people to participate in the research with agency required that I situate them as the expert, not me.

These case studies also involved ethnographic research into the making process, and the purpose of this research was to understand the dramaturgical intentions of the performance.[2] Using open-ended interviews, I was interested in the discussions that emerged from these research participants about what they thought their performances meant. I was able to understand more about how young people engaged with theatre, and how their experience of the digital culture did translate to their experience of theatre as an act of reception. The application of 'digital' themes, content and aesthetics was understood by the audiences, and they easily correlated experiences of the digital culture to those in the theatrical. Methodologically, I understand that knowledge generation in theatre is multifaceted, and that—as an audience member—knowledge generation became integral to my understanding of those works.

This was the key reflection I took to my third case study, *As If No-One is Watching*. From one of my earlier case studies, an interest in how the smartphone was used in performances emerged, and the smartphone was a central delivery device in *As If No-One is Watching*. I watched two different seasons of the show, and then interviewed the director and my other participant had performed in the first season and audienced the second. The focus of these interviews was on the mechanism of technology, specifically how the smartphone was intended to operate, operated in performance, and then was received by my second participant. I then reflected on the outcomes of these interviews considering my personal experience of the performance. The first time I saw the

---

2   By dramaturgical intentions, I mean understanding the underlying narrative and story of the performance within the context of which it was made. What choice the creators made to realise the text according to the vision of the director, and why those choices were made. From this I would be able to determine the thematic intents of the performance, and then be able to assess if those intents were realised in a way that the audiences were able to receive them.

performance, ideas about young people, theatre and the digital culture were still emergent. The second time these ideas had begun to coalesce, and I was able to reflect on those thoughts and themes as I watched the performance, and then interviewed the director and performer/spectator. The central conclusion I drew from this research was that the networks we experience in our social and digital lives, can be implicated, and replicated in our theatrical experiences. As such, the individuality of being connected over a network is more collaborative, than the solo collectivity of being an audience member. The presence of smartphones in the way that we engage with theatre also became central to my thinking around how young people's engagement with the digital culture translates into the theatrical.

My final case study, *Body of Knowledge*, occurred in the later stages of the third year of my PhD. After spending a lot of the third year wondering what the final performance would be, I came across a performance that involved young people performing for adults, using smartphones and the vehicle of performance. Until this point, I had methodologically applied empirical participatory modes of engagement influenced by an AR paradigm. I had engaged with makers and audiences of three different theatre productions, and I was confident that the knowledge I had assembled was co-generated. In this final case study, I was predominantly interested in my experience of the performance as a spectator, and so did not specifically engage with makers or other audience members. There were no workshops, surveys or interviews like the other case studies, just casual post-performance conversation at the bar and my own personal reflection and performance analysis. I was the researcher-participant. Subsequently, I was somewhat surprised when I realised that AR knowledge generation's principles had guided my analysis and could therefore be applied to how meaning is made in theatre, but also how knowledge generation, living systems and cyclic research permeate an AR project.

Enduring reflexivity: Action research in a doctoral theatre studies project

My methodological approach to the fourth case study indicated to me that while I was not actively engaging with the paradigm as an explicit methodology, my thinking around areas including knowledge generation, living systems and cycles have influenced the ways I approached my research design and analysis. The methods I wielded in my case studies reflected and built on those that I had deployed earlier. I positioned the people I spoke to as participants, not subjects in the research, and built my knowledge of the performances I considered based on the expert opinion of the makers. Where possible I named the research participants in the research, affording them ownership of what they said. I offered them the opportunity to read and listen to my research outputs, allowing them to offer feedback and corrections. Arguably, my research is participatory enquiry, influenced by the principles of an AR paradigm. While I did not see myself as an active audience or participant engaging in my final case study, on reflection, and in conversation with the ALARj (Action Research Action Learning Journal) peer reviewer, I see that in the act of identifying myself as a participant I was constituted as an active, participatory audience member. In asking generating questions based on my prior knowledge of the performance, I was replicating the act of generating post-performance questions for audience members. In answering them immediately after the performance, I was able to trace the effect of performance in the same way that I traced the effect of *Hungry Ghosts*. Overall, I was able to trace AR knowledge generation principles through my performance analysis and apply the same tenants of AR to my role as the audience that I had to those of my young participants in the other case studies.

When reflecting on my PhD project overall, it becomes clear to me that it is difficult to actively deploy a developed AR approach given the changing culture of an Australian PhD program. While the tenets of AR are more widely accepted – for example, the course, and subsequent workshop that this special issue emerged from, with a large cohort of students over several years — our programs have become time limited. I found there was no way to effectively deploy a cyclic, reflective, longitudinal process over a three-to-four-year PhD program. While I was able to revisit several participants multiple times in the data collection phase, and they had access to the findings, there was no time to engage them in the data analysis phase. Regardless, making the decision to sidestep the longitudinal aspect of my project, while adhering to the philosophical approach to research that AR brings was freeing, and my research was richer for that experience.

What I have taken from the relationship that AR has with my PhD is that applying a truly AR paradigm means that the research must come from the community. Yes, I can apply aspects of AR to my research and ensure that the participants are able to participate in the research, but until they are directing its path in a cyclic fashion of reflection and revisioning, the research can only ever be AR influenced, as my emergent understanding of knowledge generation has become. In undertaking a similar approach with audiences and theatre makers in the future, I would ask the theatre makers to direct the line of enquiry more clearly; what do they want to know about the audience experience. I would also ensure that the participants have the option of being named and therefore reflected in the research outcomes. Reflecting on the overarching AR journey I have made, I would also approach the project understanding that an AR paradigm, with the tenets of knowledge generation, living systems and cyclic research is at the heart of how I try to understand audiences.

Useful links

- https://www.mtc.com.au/plays-and-tickets/whats-on/production-archive/2015-2019/season-2018/hungry-ghosts/
- https://witnessperformance.com/hungry-ghosts-the-plane-that-never-crashes-because-it-is-always-crashing/

- https://blogs.unimelb.edu.au/creative-convergence/case-studies/
- https://performancespace.com.au/program/body-of-knowledge-by-samara-hersch/
- https://rundog.art/body-of-knowledge-samara-hersch/
- https://www.stagewhispers.com.au/reviews/if-no-one-watching
- https://www.nothingeverhappensinbrisbane.com/review-archive/2018/09/29/as-if-no-one-is-watching-vulcana-womens-circus-and-waw-dance
- https://www.artshub.com.au/news/reviews/review-as-if-no-one-is-watching-brisbane-powerhouse-256575-2360962/

Acknowledgements

Research for two of the case studies, *Hungry Ghosts* by Melbourne Theatre Company and *Take Over: A Digital World* by Geelong Arts Centre was undertaken as part of the Australian Research Council Linkage Project *Creative Convergence: Enhancing Impact in Regional Theatre for Young People*, reference number LP160100047. The research methodology used for these projects drew on that developed by Rachel Fensham, Paul Rae, Meg Upton and Jennifer Beckett.[3]

References

Ashman, Emina, Jing-Xuan Chan, Petra Kalive and Bernard Sam. (2018). Personal interview, interviewed by Paul Rae and Abbie Trott, Melbourne, 27 April 2018.

---

3 Rachel Fensham and Megan Upton discuss the post-performance methodology developed throughout the project in their journal article in RIDE: The Journal of Applied Theatre and Performance (2022).

Ashman, Emina, Jing-Xuan Chan and Bernard Sam. (2018). Personal interview, interviewed by Paul Rae and Abbie Trott, Wangaratta, 5 June 2018.

Bakhshi, H & Throsby, D 2010, *Culture of innovation: An economic analysis of innovation in arts and cultural organisations*, NESTA. https://nesta.org.uk/report/culture-of-innovation/ [Accessed 3 April 2017].

Balme, C, 2014, *The Theatrical Public Sphere* Cambridge UP, Cambridge.

Barker, M, 2003, 'Crash, theatre audiences, and the idea of "liveness"', *Studies in Theatre and Performance*, vol. 23, no. 1, pp. 21-39. DOI: 10.1386/stap.23.1.21/0.

Baxter, L, O'Reilly, D & Carnegie, E, 2013, 'Innovative methods of enquiry in arts engagement', in Radbourne, J, Glow, H & Johanson, K (eds.) *The audience experience: a critical analysis of audiences in the performing arts*, Intellect Books, Bristol, pp. 113-128.

Brown, AS & Novak, JL, 2007, *Assessing the intrinsic impacts of live performance*, Wolf Brown. https://culturehive.co.uk/wp-content/uploads/2013/04/ImpactStudyFinalVersion.pdf [Accessed 3 April 2017].

Conquergood, D, 1988, 'Health theatre in a Hmong refugee camp: performance, communication, and culture', *TDR*, vol. 32, no. 3, pp. 174-208. https://doi.org/10.2307/1145914 www.jstor.org/stable/1145914 [Accessed 21 June 2017].

Conquergood, D, 1991, Rethinking ethnography: Towards a critical cultural politics. *Communication Monographs*, vol. 58, no. 2, pp. 179-194. DOI: 10.1080/03637759109376222 https://web.a.ebscohost.com/ehost/detail/detail?vid=2&sid=7e985c6a-5f50-448c-a9c7-7911f2de763c%40sessionmgr4006&bdata=JkF1dGhUeXBlPXNzbyZzaXRlPWVob3N0LWxpdmU%3d#AN=9107290447&db=ufh [Accessed 21 June 2017].

Conquergood, D, 2002, 'Performance studies: interventions and radical research', *TDR* vol. 46, no. 2, pp. 145-156. www.jstor.org/stable/1146965 [Accessed 7 April 2017].

Cultural Development Network. (2016). *A schema of measureable outcomes for cultural engagement*. Cultural Development Network. Available from: https://culturaldevelopment.net.au/planning/measurable-outcomes-of-cultural-development-activity-within-the-six-domains/ [Accessed 22 September 2017].

Dorsey, J, 2016, *Gen Z: Tech disruption*. The Centre for Generational Kinetics. Available from: https://genhq.com/wp-content/uploads/2017/01/Research-White-Paper-Gen-Z-Tech-Disruption-c-2016-Center-for-Generational-Kinetics.pdf [Accessed 8 June 2020].

Fensham, R & Upton M, 2022, 'Post-performance methodologies: The value of memory for theatre with young people', *Research in Drama Education: The Journal of Applied Theatre and Performance* DOI: 10.1080/13569783.2022.2097864. Accessed 25 July 2022].

Foreman-Wernet, L & Dervin, B, 2013, 'In the context of their lives: how audience members make sense of performing arts experiences', in Radbourne, J, Glow, H & Johanson, K (eds.), *The audience experience: a critical analysis of audiences in the performing arts*, Intellect Books, Bristol.

Freshwater, H, 2009, *Theatre & audience* Palgrave Macmillan, Basingstoke, UK.

Frieze, J, 2017, 'Introduction', in Frieze, J (ed.), *Reframing immersive theatre: the politics and pragmatics of participatory performance*, Palgrave Macmillan.

Gallagher, K, 2014, *Why theatre matters: urban youth, engagement, and a pedagogy of the real*, University of Toronto Press

Geelong Lutheran College. (2018). Research Workshop, facilitated by Abbie Trott, Geelong, 6 Aug. 2018.

Geelong Lutheran College. (2019). Research Workshop, facilitated by Abbie Trott and Kelly Clifford, Geelong, 10 Sept. 2019.

Greenwood, D & Levin, M, 2007, *Introduction to Action Research: social research for social change*. Sage Publications, Thousand Oaks..

Heim, C, 2017, *Audience as performer: the changing role of theatre audiences in the twenty-first century*, Routledge, Oxam. Available from: https://www.routledge.com/Audience-as-Performer-The-changing-role-of-theatre-audiences-in-the-twenty-first/Heim/p/book/9781138796928 [Accessed 24 May 2018].

Kershaw, B, 2001, 'Oh for unruly audiences! Or, patterns of participation in twentieth-century theatre', *Modern drama*. Vol. 44, no. 2, DOI: 10.1353/mdr.2001.0029 [Accessed 18 March 2017].

Lewis, WW & Johnson, S, 2017, 'Theatrical reception and the formation of twenty-first century perception: A case study for the igeneration', *Theatre topics*, vol. 27, no. 2, pp. 123-136. DOI: 10.1353/tt.2017.0024. [Accessed 19 April 2021].

McConachie, BA, 2008, *Engaging Audiences: a cognitive approach to spectating in the theatre*, Palgrave Macmillan, Basingstoke. https://link.springer.com/book/10.1057/9780230617025 [Accessed 27 February 2017].

Nelson, R, 2006, 'Practice-as-research and the problem of knowledge', *Performance Research,* vol. 11, 4, pp. 105-116. DOI: 10.1080/13528160701363556 [Accessed 18 November 2019].

Peck, C, 2015, 'I want to play to': why todays youth are resisting the rules of the theatre. *Theatre Symposium*, vol. 23, pp. 124-136. DOI: 10.1353/tsy.2015.0004 [Accessed 30 November 2016].

Radbourne, J, Glow, H & Johanson, K, 2013, 'Knowing and measuring the audience experience', in Radbourne, J, Glow, H & Johanson, K (eds.), *The audience experience: a critical analysis of audiences in the performing arts*, Intellect Books, Bristol.

Reason, M, 2006, 'Young audiences and live theatre, part 1: methods, participation and memory in audience research', *Studies in Theatre and Performance,* vol. 26, no. 2, pp. 129-145. DOI: 10.1386/stap.26.2.129/1

Reason, M, 2010, *The young audience: exploring and enhancing children's experiences of theatre*, Trentham Books Ltd, Stoke on Trent.

Sedgman, K, 2016, *Locating the audience: how people found value in National Theatre Wales*, Intellect Books, Bristol.

Sedgman, K, 2019, 'On rigour in theatre audience research', *Contemporary Theatre Review*, vol. 29, no. 4, pp. 462-479. DOI: 10.1080/10486801.2019.1657424 [Accessed 7 May 2020].

Thompson, J, 2012, *Applied Theatre: Bewilderment and Beyond* (4th edn), Peter Lang, Oxford. Available from: https://www.peterlang.com/document/1043629 [Accessed 14 November 2017].

Tulloch, J, 2000, 'Approaching theatre audiences: Active school students and commoditised high culture', *Contemporary Theatre Review*, vol. 10, no. 2, pp. 85-104. DOI: 10.1080/10486800008568588 [Accessed 1 March 2017].

Twenge, J, 2017, *iGen: Why today's super-connected kids are growing up less rebellious, more tolerant, less happy--and completely unprepared for adulthood--and what that means for the rest of us*, Atria Books, New York. https://web.b.ebscohost.com.ezproxy.library.uq.edu.au/ehost/detail/detail?vid=0&sid=d1bff703-d83b-4790-aca4-4452a383d33a%40pdc-v-sessmgr05&bdata=JnNpdGU9ZWhvc3QtbGl2ZQ%3d%3d#AN=1955950&db=nlebk [Accessed 8 June 2020].

Wadsworth, Y, 2011, *Building in research and evaluation: human enquiry for living systems* (1st edn), Action Research Press, Hawthorne, Vic.

Biography

Interested in examining the reception of digital bodies, Dr Abbie Victoria Trott researches post-digital theatre with audiences. Teaching theatre and performance at a tertiary level since 2014, she is an experienced stage and production manager across community theatre, circus, and multimedia performance. Abbie lectures in Drama at the University of Queensland and is working at Deakin University as a Research Fellow on a project about audience diversity and organisational change in Australian arts and culture.

An HRM student's search for relevancy

Edward Hyatt

Abstract

The following paper details how my PhD journey and dissertation reflected the ethos of Action Research (AR), a research paradigm that emphasizes collaborative knowledge creation between researchers and community members with the aim of solving practical, real-world problems. It attempts to make the case that adopting AR principles can help ameliorate a crisis of relevancy in business research and help to bridge a researcher-practitioner divide in the human resource management (HRM) field. I first provide context with a brief background about myself and my understanding of the core principles of AR. This is followed by a broader discussion of a perceived crisis in the practical relevancy of management research and psychometric-focused HRM research. I then return to personal matters to discuss the evolution of my dissertation and how it might qualify as participatory research in spirit if not entirely in practice, especially in its use of descriptive phenomenology. I highlight specific challenges with thesis writing and working with supervisors and conclude by reflecting on how my individual challenges may be generalizable to other PhD students seeking to produce impactful and practicable research.

Key words: PhD journey, action research, researcher-practitioner divide, human resource management, descriptive phenomenology

What is known about the topic?

In both business literature and HRM research there is a well-recognized disconnect between the worlds of academics and practitioners, often dubbed the researcher-practitioner divide. This is exemplified by the low uptake among practitioners of structured job interviews, which research has firmly established are superior to their unstructured counterparts. Regarding AR, there are some readily recognizable principles (e.g., co-generation of knowledge, a problem-solving focus) that would appear to offer great promise for bridging that researcher-practitioner divide.

What does this paper add?

The central purpose of the paper is to present the challenges and rewards of pursuing AR-inspired research as a PhD student. More broadly, it seeks to make the case that adopting an AR mindset can help ameliorate the crisis of relevancy in business research. The paper also seeks to show how descriptive phenomenology is closely aligned in spirit with AR practices.

Who will benefit from its content?

Primarily future PhD students, especially those wanting to adopt an AR mindset within the confines of a research paradigm that might not support certain AR principles.

What is the relevance to AL and AR scholars and practitioners?

The relevance to fledgling AR scholars is in the lessons learned that are specific to overcoming challenges experienced by adopting an AR mindset. More senior AR scholars might be exposed for the first time to descriptive phenomenology as a research method

Received   January 2022   Reviewed May 2022   Published November 2022

Introduction

When Professor Akihiro Ogawa first started the conversation about a special issue centered on PhD student experiences using action research (AR) for their dissertation, I was hesitant to participate in the endeavor. I hesitated because although I had initially wanted to use AR as the primary methodology for my recently completed thesis, I was unable to ultimately do so. I could not call my thesis an AR project in the strictest sense of the word, which made me question the value of including my voice alongside other truer AR projects. For instance, my research topic was not emancipatory in any fashion. I dealt with the rather vanilla issue of managerial resistance to using structured job interviews as a means of hiring people into a company. I could not entertain multiple, iterative cycles of research or interventions along with community involvement that are often a hallmark of AR projects. If anything, my research topic was more identifiable as being aligned with a Northern, Western tradition of action research as it involved an industrial-type problem to be investigated rather than being an example of the Southern liberationist form of participatory action research (see Greenwood and Levin 2007 for a primer on the two major strands of AR).

However, both who I am as a researcher and the spirit with which I approached my thesis is consistent with principles that I believe underpin the AR philosophy. My thesis was an expression of my desire to solve real problems, to interact directly with end-users (or in academic parlance, "practitioners"), and to contribute both to action (being practicable) and research (expanding theoretical knowledge). I eventually used a qualitative methodology called descriptive phenomenology (Husserl 1973), which relies on in-depth interviews to develop a holistic understanding of a human subject's unique perspective of a phenomenon (Moustakas 1994). This served my dissertation purposes quite nicely and is philosophically aligned with the AR sentiment that people "have detailed, complex, and valuable knowledge about their lives, environments, and goals" (Greenwood & Levin 2007, p. 103). I believe this type of mindset needs to be adopted more often in order to ameliorate a crisis of relevancy in business management research, especially in the field of HRM and selection research.

So, I agreed to continue with this project and hopefully I can make a practical and meaningful contribution to future graduate students by reflecting on my PhD experience and how it evolved within the domain of management research. In order to properly flesh out my journey and the intended contribution of the paper, some context first needs to be provided. I will give a brief background about myself, followed by my personal understanding of what is AR. I will then shift gears and briefly discuss a perceived crisis in the practical relevancy of management research, before going more in depth on how much of HRM research is myopically focused on psychometric issues to the detriment of developing a better understanding of practitioners as potential users of research output. This broader literary context provided the milieu and motivation for the critical thinking that helped shaped my research experiences and thesis direction. I will then return to personal matters to discuss the evolution of my dissertation and add some personal reflections of how my journey may reflect the broader issues already discussed. In conclusion, I hope to produce something both personal and generalizable enough to be useful to others embarking on an AR-inspired PhD journey.

Myself

In my brief time as a researcher, I have come to identify myself as a pragmatist in the vein of William James, Charles Sanders Peirce, and John Dewey. I frequently think in an abductive manner in both informal and scientific settings. Abductive reasoning is a term originally coined by the philosopher Charles S. Peirce (1931-1958) to describe a form of reasoning distinct from deductive and inductive reasoning, the former usually associated with quantitative research and the latter with qualitative research. Abduction is an inferential logic used to offer plausible or likely explanations in order to make sense of surprising facts or puzzling observations (Thagard & Shelley 1997). Abductive reasoning begins with an observation or set of observations and, based on clues contained therein, gives rise to "speculations, conjectures, and assessments of plausibility" (Weick 2005, p. 433) to explain how that phenomenon may exist. Abduction is not meant to provide the same level of certainty or positive verification as deductive logic, but instead allows the researcher to offer a credible explanation that *may* account for the observation. I often use this type of thinking when dealing with a problem or curious set of facts, especially at the outset when I have more limited information at hand.

This type of thinking lends itself well to my ontological positioning as a researcher. Ontology can be described as the "theory of being" (Delanty & Strydom 2003); in other words, an account of what is the nature of reality. I adopt an ontological position that places me in a conceptual middle ground between the polar extremes of objectivism and constructivism, something akin to being a critical realist (Bhaskar 2013; Sayer 2010). On the one hand I believe there is a physical world that exists independently of human awareness and that is inherently devoid of meaning. However, I also believe there is a social reality composed of interpretations, concepts, and meaning that is uniquely constructed through human language and social interaction. Like Chalmers (2013) I believe that a "single, unique, physical world exists independently of observers… But it does not follow from this that they [individuals] have identical perceptual experiences" (p. 9). I take it as a given that these dual realities are inexorably enmeshed with one another and that what humans conceive as real is a unique combination of external reality and linguistic meaning. Consequently, our social reality can only be fully known through a combination of measurement and direct observations of reality *and* meaningful inferences and interpretations of human thought and activity.

Given my way of thinking and ontological alignment, I am fundamentally less interested in what is "true" (as in a single, distinct, measurable reality) than in what is "accurate" or "what works" per the context, the persons involved, the time, and other features of social reality. This reflects my core belief that the true value of science "resides in its potential to develop conditions (material and symbolic) that are beneficial to human beings" (Alvesson & Willmott 1992, p. 436), which necessarily means making a contribution beyond the realm of academia. It simply does not make sense to me to investigate phenomena unless the resulting knowledge is likely to be actionable, solve a problem, or somehow improve the human condition. Conducting research for theory's sake or publishing solely for the sake of my own career seems like a missed opportunity to make the most of a privileged position. As it happens, this sentiment is highly commensurate with AR, which is why I was immediately attracted to it as a method of enquiry and research.

My understanding of action research

Action research is an investigative approach that seeks to foster conscientious learning and decision-making through experiential inquiry and continuous cycles of action and reflection. It has been described as "a research practice with a social change agenda" (Greenwood & Levin 2007, p. 4) that advocates not just studying social problems from a distance but also trying to actively resolve them. It could be said that this type of research approach is one where "the location of knowledge is secondary to the larger question of the problem that needs to be solved" (Creswell & Miller 1997, p. 39). Similar to my own personal stance, AR considers truth to be provisional, or the best reasoned knowledge, to be gained through a co-generative process between academic researchers and practitioners. The interest in solving practical, real-world problems means that the final product of a research undertaking should be beneficial to human beings, which frankly seems to be a goal absent from many mainstream research philosophies that value objective "truths" or embody a theory-for-theory's-sake mentality.

Action research is a co-generative knowledge creation process whereby a researcher works collaboratively with community members to solve real-time problems (Lincoln et al. 2011). This form of research necessitates a different, more democratic relationship between professional researchers and practitioners than is usually adopted in research endeavours. First, the practitioner identifies the problem to be solved, not the academic. Second, practitioners are active participants throughout the research process, offering expert local knowledge and validation of the ongoing research effort because they are the stakeholders driven to act in their environments. As described by Greenwood and Levin (2007), "good AR practice is to design and sustain a process in which important reflections can emerge through communication and some good practical problem solving can be done in as inclusive and fair a way as possible" (p. 113). This problem-solving approach often requires multiple cycles of planning, implementation, data collection and active observation, reflection, and evolution of the overall project. In short, AR seeks to study social problems and generate useful knowledge in a highly democratized fashion with actors outside of academia.

But AR is more than consulting with a heart; it is still research, after all. It also contains the conventional scholarly duty to contribute to theory; the process still needs to generate implications beyond those required for immediate action (Eden & Huxham, 1996). This dual imperative to produce workable solutions to real-world problems *and* extend theory distinguishes action research from consultancy and other forms of scholarly research (McKay & Marshall 2001). The simultaneous and distinct interests in problem-solving and research (see Figure 1) can be an additional burden for an AR researcher that do not normally exist for other researchers, especially those in conventional academic knowledge systems which primarily incentivize scholarly publications for other academics. High levels of practitioner engagement, power-sharing, and longitudinal research designs mean that AR is not for the faint of heart, yet it holds great promise for generating more relevant and impactful research, especially in fields of research that exhibit researcher-practitioner divides.

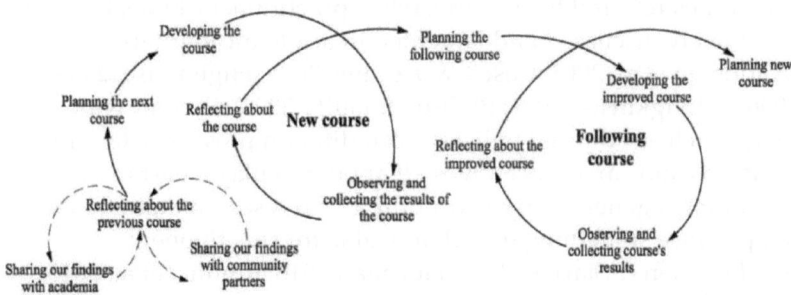

Figure 1: The action research cycles led to continuous course improvement
(Source: do Amaral, J. A. A. and Okazaki E. 2016)

Business literature: Crisis of (lack of) relevancy

It is important to consider the literary environment in which I believe I operated during my PhD program in order to better understand my own experience. Beyond its intrinsic qualities, I believe AR is uniquely suited to address a lack of perceived relevancy of business research in general and HRM research in particular.

My primary area of research is focused on employee selection practices (e.g., interviews), which is a subfield of human resource management, itself embedded within the broader umbrella of business management and organizational research. Despite the obvious value that evidence-based research can bring to the business world, little of the research produced by business schools has been perceived as actionable or deemed relevant by the management profession (Beer 2001; Bennis & O'Toole 2005; Ghoshal 2005; Mintzberg & Gosling 2002). There is a broad disconnect between the worlds of academics and practitioners as organizations regularly fail to implement research findings in their workplaces (Dipboye 2007; Rynes et al. 2002). This trend, sometimes referred to as a researcher-practitioner divide, has been extensively discussed and lamented in academic journals (Anderson et al. 2001; Cascio & Aguinis 2008; Highhouse, 2008a, 2008b; Hodgkinson 2011; Phillips & Gully 2008). System-wide issues such as editorial policies, accreditation pressures, tenure decisions and career incentives all appear to play a part in sustaining a generally narrow approach to research that often does not produce something useful or usable for practitioners (Hodgkinson & Starkey 2011; Hoffman 2016; Romme et al. 2015; Starbuck 2006).

Concerns regarding the perceived lack of research relevancy have become more important to academicians, practitioners, and policymakers in the last several decades (Hodgkinson & Starkey 2011). For instance, the question of whether academic research "mattered" was a centerpiece of Hambrick's 1993 Academy of Management Presidential Address (Hambrick 1994). A few research approaches have been suggested for making research and knowledge production more relevant to practice, including action science (Beer 2001), design science (Hodgkinson & Starkey 2011), and mode 2 knowledge-production (Nowotny et al. 2001). All these approaches are similar in that they encourage academics to frame research problems in the context of real-world application. They share a "pragmatic concern for effectiveness ('does it work?') rather than 'truth' ('is it true?') as a guiding research principle" (Hodgkinson & Starkey 2011, p. 363). As well, Pettigrew (1997, 2001) made the seminal argument that good management research needs to fulfil the 'double hurdles' of rigor and relevance, mirroring the dual imperatives of action research (McKay & Marshall 2001).

One feature of action research seems to be particularly relevant for addressing the lack of practical relevancy of academic research: that is, the extensive involvement and collaboration between practitioners and academics. Concerns regarding the practical impact of business research have prompted a chorus of calls for higher involvement between the two worlds of academia and practice (e.g., Pettigrew 2001; Romme et al. 2015; Rynes 2007; Tranfield & Starkey 1998; Van de Ven & Johnson 2006). Tranfield and Starkey (1998) advocated for management research activity where "the problems addressed by management research should grow out of the interaction between the world of practice and the world of theory, rather than out of either one alone" (p. 353). Several researchers have explicitly called this "engaged scholarship" (Hoffman 2016; Hughes et al. 2011; Van de Ven & Johnson 2006) which again bears striking resemblance to action research. This higher level of interaction between academics and practitioners remains largely missing in business literature; hence the calls for academics to engage more directly with practitioners to identify research problems and design studies that make a difference in the world outside of ivory towers (Hodgkinson & Starkey 2011; Hoffman 2016; Romme et al. 2015; Rynes et al. 2001; Shapiro & Kirkman 2018).

The researcher-practitioner divide is often characterized in scholarly texts as a one-way knowledge transfer issue from the expert researcher to the stubborn layman (e.g., Tranfield et al. 2003). However, it might be more constructive to approach it as a knowledge creation problem (Rynes et al. 2001; Starkey & Madan 2001). Knowledge creation in this sense is a social process that requires both parties to interact and codify their understanding of one another. After all, one of the key conditions for an effective exchange of knowledge is a mutual understanding between parties of each other's underlying assumptions and motivations (Guzman & Wilson 2005). It is here again that one can see the potential value in adopting a participatory action research approach since it is a democratizing process that enables academics and practitioners to understand one another as equally contributing partners. Of course, one can also see how this approach challenges the notion of researchers as the sole authoritative source of knowledge, and it can be hard to implement this type of research in hierarchical organizations that do not necessarily espouse democracy within their own ranks. Nevertheless, higher levels of engagement should improve the perceived usefulness of academic research if only because it would force the two parties to interact, compare perceptions, examine preconceived notions, reveal motivations, and ultimately generate greater overall understanding of each other.

HRM field: Psychometric heavy

As noted in the previous section, there is a well-recognized disconnect between the worlds of business academics and practitioners. In fact, the widest gap between research and practice is arguably in my field of study, the area of employee selection (Rousseau & Barends 2011; Rynes 2012). Practitioners have reportedly found many of the tools and practices supported by HRM research to be cumbersome, irrelevant to their own work situation, or simply disappointing (Anderson 2005; Yates et al. 2003). Reflecting the business school rationale for the existence of a researcher-practitioner divide, this situation is usually attributed to practitioner stubbornness, bias, or lack of knowledge. However, while it is not usually spelled out in exact terms, much selection research is conducted along a set of assumptions and values that are more aligned with a researcher's mindset than a practitioner's mindset.

Most existing HRM literature is preoccupied with psychometric concerns about reliability and validity, themselves based on traditional managerial concerns for measuring and fitting individuals to organizational purpose (Weiss & Rupp 2011). HRM research has largely considered individual differences in predicting future job performance to be the primary area of research interest (Guion 1998; Harley 2015). Consequently, the focus has been on how to improve the psychometric qualities of selection methods like job interviews, especially their criterion validity as it relates to predicting individual job performance (Cascio & Aguinis 2008; Huffcutt & Culbertson 2011). This perspective does not generally concern itself with other topics such as implementation and user acceptance that are probably of high interest to practitioners working in organizational realities (Muchinsky 2004).

As a corollary, most academic efforts at encouraging the use of evidence-based practices focus on how to better communicate the value of statistical validity (e.g., Highhouse et al. 2017; Rauschenberger & Schmidt 1987; Tranfield et al. 2003; Zhang et al. 2018). This approach reflects an implicit assumption that "the truth will prevail if it is appropriately presented" (Skarlicki et al. 1996, p. 17). This, however, clearly does not work. A damning observation that has been reinforced by multiple surveys over time, generalized across types of organization, applicant groups, and countries, is that the use of selection methods is inversely proportionate to their reliability and validity (Shackleton & Newell 1994; Zibarras & Woods 2010). For example, in a study examining practitioner rationale for using selection procedures, non-economic predictors like applicant reactions and organizational self-promotion were more strongly associated with use of a selection procedure than was perceived predictive validity (König et al. 2010). Rather than wondering why this pattern persists and investigating it closer, academics have characterized the preference for less valid selection methods as practitioners being stubborn or unwilling to learn (Highhouse 2008b).

A possibility that does not appear to be heavily considered in traditional HRM research is that the very separate worlds of academic and practitioner, as described by Rynes et al. (2007), might exist because what gets frequently researched is missing key features of working life from a practitioner perspective. Little work has been done on user perspectives and how to make decision aids more acceptable or attractive for use in organizational settings (Diab et al. 2011; Kuncel 2008). This is almost certainly the case because user perspectives are simply not deemed worthy of investigation. There are important areas such as "managers' real *purpose* and their *tacit knowledge* about why and how they do what they do" (Beer 2001, p. 60 [emphasis in original]) that remain understudied because of lack of scholarly interest.

I came to believe early in my dissertation process that some of the assumptions about the purpose and nature of job interviews had become so taken for granted in selection research that they were no longer seriously discussed or questioned (Argyris & Schön 1996). As Rynes et al. (2001) have suggested, these types of unquestioned premises may be at the heart of the lack of perceived relevancy and credibility of research findings to practitioners. It appeared to me that the low uptake of structured interviews was not merely a knowledge-transfer issue, and that a more fundamental understanding of the practitioner worldview would behoove researchers interested in producing relevant selection research. All this thinking and observation helped to shape my own PhD journey and eventual thesis.

My PhD journey

The following section details my own dissertation journey. As required of any PhD student I began my exploration by reading as much as possible on my area of interest, in this case job interviews as a means of selecting people to work in organizations. My initial canvassing of the literature resulted in background knowledge and two key observations.

The first observation was that meta-analytic research has consistently demonstrated that structured interviews are more valid than their unstructured counterparts for predicting job performance (Huffcutt & Arthur 1994; Huffcutt et al. 2014). This evidence is rather conclusive; structured interviews are more valid, and researchers continue to generate evidence discouraging the use of unstructured interviews (e.g., Dana et al. 2013). However, and this is the second observation, despite the strong evidence for the benefits of interview structure from a psychometric perspective, practitioners still prefer the unstructured interview. For more than a century the unstructured interview has remained more popular and widely used in practice than the research-supported structured interview (Buckley et al. 2000; Rynes 2012; Wonderlic 1937). A number of industry surveys over time have continued to provide evidence of the low uptake of structured interviews among organizations (Ryan et al. 1999; Rynes et al. 2002), including ones conducted outside the U.S. (Sanders et al. 2008). This situation perfectly embodies the researcher-practitioner divide discussed in the earlier section.

I was struck by the low levels of uptake of an obviously superior job interview format that had so much scholarly evidence for achieving the business goal of hiring effective employees. In my mind, I kept returning to the question of *why* practitioners would not use the structured interview. The simplest explanation I devised was that maybe they were not using job interviews for the purposes that academics assumed they should be using them. Perhaps they were using interviews to accomplish something other than identifying the most technically qualified person, and maybe structured interviews were not as effective in achieving those other aims. Besides hypothesizing some of those other potential outcomes of interest to practitioners, I also began to reflect on the implicit assumptions behind descriptions of so-called practitioner resistance in the existing research. I began to consider the possibility that it was the research that was missing something important from the interviewer's perspective, not the other way around. After some initial informal conversations with recruiters and HRM professionals I came to believe that practitioner "resistance" was not simply a matter of ineffective communication, ignorance, or unwillingness on the practitioners' part. Rather, I could sense that there was a lack of understanding on academics' part of the practitioner worldview and concerns. I began to see how the field of HRM research had likely not done enough to reflexively consider that the researcher-practitioner divide may be partly due to shortcomings on the academic side (cf. Rynes 2012).

So, in abductive fashion, my take on the situation could be summarized as follows: (a) there is a well-substantiated "better" way to conduct job interviews for predicting an applicant's future work performance, (b) there is a persistent situation whereby practitioners do not use structured interviews despite the research, (c) this divide is usually attributed to practitioner resistance or some inability for academics to effectively communicate the value of structured interviews. However, a plausible alternative explanation is (1a) this gap might exist at least in part because of a lack of understanding on academia's part of practitioners' worldviews, and (1b) this lack of understanding likely persists because the dominant HRM research paradigm prioritizes psychometric concerns over practical ones. It was not just the question of why practitioners avoid using unstructured interviews that drove my interest; it was also the realization that the broader researcher-practitioner divide might exist because of an academic mindset. I figured this necessitated adopting an alternative perspective to develop meaningful insights in an otherwise well-researched area.

I initially designed a series of experiments to test a plausible explanation for the low practitioner uptake of structured interviews. However, my supervisors felt there was not enough existing literary support to justify the investigation. I spent the better part of my first year trying to convince them that the topic was valuable precisely because the lack of existing evidence meant it challenged assumptions embedded in the field. My personal impression of most of these conversations with my supervisors was that *I* could not understand how *they* could not understand that this was an important topic worth investigating. I tried making my case in various ways: using hypothetical examples, providing written summaries, responding point-by-point to early feedback, making presentations, using sample quotes from preliminary interviews, bringing in guest speakers to classes, presenting anecdotal evidence from my own past work experience, and so forth. I believe that my arguments were ultimately unpersuasive; for much of that time I felt like I was trying to explain water to a fish. I believe they were immersed in an ontological perspective of HRM research that I had not fully taken for granted, and that I could not sell them on the value of the project (as was my responsibility as the student). I have come to appreciate their perspective far more with some time and distance and realize that they likely considered the project innately valuable but simply felt that it would be hard to publish without what is generally considered the more valuable outcomes of interest in the field (e.g., company performance). Regardless, I feel like we talked past each other for most of this time, with little productive output to reflect the frustration that all of us were undoubtedly feeling.

The result of this impasse was that I abandoned the initial experimental study design and instead focused all my energies on a single, qualitative study. This was done in order to accomplish the theory building that was deemed necessary to justify my original desired research efforts. In essence, I needed to treat my thesis as a purely exploratory endeavour to substantiate a more detailed investigation in the future. I was disappointed, but I also felt after a year of circular conversations that a basic understanding and external validation of the intrinsic value of the topic had to precede any other substantive work. I was also comforted to pursue a solely qualitative effort because I could not risk the possibility that a fatal flaw in my experimental study design would escape notice simply because my supervisors and I were not on the same page. Although I was initially disappointed, two important outcomes resulted from the overhaul in my thesis direction.

First, my supervisors' insistence for literary support led me to discover an already explicit critique of the HRM system very similar to my own thoughts. According to Herriot (1992, 1993), two professional sub-cultures with different values and assumptions co-exist within the academic world of selection research. The more dominant psychometric sub-culture is fundamentally managerialist and views the job interview as a type of psychometric instrument, hence the concern with criterion and predictive validity. The social sub-culture values both individuals and organizations as clients and regards the job interview as a socially interactive process influenced by practical concerns. Herriot conjectured that the social sub-culture might be more well suited for developing research that appeals to organizational clients; by extension, this would likely help ameliorate the low uptake of research findings in practice. Here was an apt conceptualization for what I had come to intuitively perceive in selection research. Herriot's (1992, 1993) conception of professional sub-cultures within selection research provided important support and clarity for my own positionality.

The second outcome was that I chose a research philosophy and design approach, descriptive phenomenology (Husserl 1973), that I believe is closely aligned in spirit with action research. As a philosophy, descriptive phenomenology values the actual, lived experiences of individuals and how they describe and interpret the world around them (Gill 2014; Moustakas 1994). As a methodology, descriptive phenomenology is concerned with establishing knowledge of a phenomenon's "essence", or core structure, by reducing human subjects' descriptions of their experiences and perceptions to the essential structures of subjective experiences (Husserl 1973). The nuances of the stories and language used by the subjects to describe a phenomenon are carefully examined to develop a holistic understanding of the subject's own perspectives on the phenomenon (Moustakas 1994). The express purpose of descriptive phenomenology is to gain an understanding of a phenomenon from the perspective of the social actors who experience it. The use of descriptive methodology represented a purposeful departure from the traditional psychometric sub-culture perspective that normally encourages arms-length survey work. Therefore, it fit quite nicely within Herriot's description of a social sub-culture and was nicely aligned with my own desires for a more democratizing and participatory form of research.

Lessons learned

I learned several important lessons during my PhD, some of which may be useful for others to hear.

First, I am grateful for the collapse of my initial study design because it allowed me to pursue a study that I felt ultimately makes a larger point of challenging the status quo thinking in selection research. It was important to me to directly address the research-practitioner gap by treating the experiences and views of interviewers as valid expressions of their reality. It bugs me that much academic research is being ignored or deemed irrelevant by the very people who should benefit the most from it. Unfair and ineffective interviews are a disservice to applicants who put their heart and soul into job searches, and bad hires are costly for everyone involved: organizations, managers, and individuals. I figured gaining direct insight into the perceptions and motivations of interviewers was an important avenue of research, but upon reflection this was not emphasized enough in the earlier study design of my dissertation. The final thesis was much closer to AR in spirit, and I found it invigorating to deal more directly with what I thought was an inherently important issue.

Second, because I had to focus on the single qualitative study, I think I was able to make a secondary methodological contribution to HRM research through my use of descriptive phenomenology. Descriptive phenomenology is uniquely suited as a means of developing a better understanding of the practitioner worldview because it acknowledges the inherent legitimacy of subjective experience and regards individuals as mindful actors able to describe their reality in a coherent and meaningful way. This sentiment towards practitioners as clients or even co-generative partners is rarely reflected in HRM research but would be strongly advocated for by action researchers. Its application in my thesis was an endorsement of more engaged relationships with practitioners, treating them as true clients and partners of our work, in order to make the academic research more relevant (Hodgkinson & Starkey 2011; Hoffman 2016; Shapiro & Kirkman 2018). I describe this contribution to another field of research here in an article dedicated to AR because I think it illustrates for future graduate students what is still possible if your initial thesis plan faces challenges from existing paradigms. I believe that my PhD research contributed to the research and practice of HRM, and now with the story of my PhD journey I hope to also contribute to the burgeoning future of AR research.

Third, I learned that participatory research can be a hard sell. I spent a long time having circular conversations with my supervisors and finding repeated comments on written drafts. The confusion was on both ends; we kept talking past each other and I do not know if a common understanding was ever achieved. I believe they were well-intentioned with their concern that not having individual performance as an outcome variable meant future difficulty publishing in HRM journals. This should be a key concern of any supervisor preparing a student for professional academic success, but it also reflects the psychometric-heavy paradigm that I believe is strangling the production of more useful research. The refrain of "you are too practical" from my supervisors and others is understandable when considering the context of the publishing game and the priorities that are incentivized by the academic system. It can be difficult to justify the worthiness of pursuing AR research devoted to resolving practical problems when the system does not reward such activity. I am grateful for everything my supervisors provided and they have big hearts, integrity, and plenty of patience; however, I am thankful for the practical boundaries of time and money that forced me to cut bait and say it now MUST get done for me to finish at all. Otherwise, I imagine I could still be having debates rather than writing articles on the subject.

Conclusion

An AR project takes a lot of time and energy; undertaking such a study is a highly ambitious affair for even the nominally long timeframe of a PhD. In my case I had to abandon my experimental study design in order to complete my PhD in my own lifetime, although I have continued to pursue those other studies as part of a larger research agenda. I am glad I undertook a study that closely resembled AR in spirit if not research design. The best thing about my PhD experience was interacting directly with practitioners and feeling like I was tapping into a user perspective that I had not seen heavily reflected in mainstream human resource literature. It was also personally gratifying to see how the people I interviewed often started to reflect on their own behavior patterns and assumptions through our conversations. This hopefully provided them some value and helped foster a greater understanding on their part of the researcher perspective.

My PhD experience made it apparent to me that the complexity of organizational life demands a healthy dose of respect from those who study it. Social scientists who are active in the field (as opposed to those ensconced in an ivory tower conducting experiments on undergraduates) are regularly faced with the difficulty that "reality backs up while it is approached by the subject who tries to understand it" (Meacham 1983, p. 130). This is something that is made much more explicit in participatory forms of research like AR that seek to make a real-world contribution in addition to theory building. I believe that more humility, a problem-solving mindset, and greater participatory efforts such as those embodied in AR are likely needed to produce a breakthrough in the relevancy of business and HRM scholarship. I think business schools may sense that forms of participatory research like AR are a way forward as well, as reflected in the increased use of the nebulous term 'impact' in the hallways and slogans of business schools. The value of participatory research for academics, practitioners, and society is such that I will continue to advocate for its use as I progress in my academic career.

References

Alvesson, M & Willmott, H 1992, 'On the idea of emancipation in management and organization studies', *The Academy of Management Review*, vol. 17, no. 3, pp. 432-464 https://doi.org/10.2307/258718.

Anderson, N 2005, 'Relationships between practice and research in personnel selection: Does the left hand know what the right is doing?', in A Evers, N Anderson & O Voskuijl (eds.), *The Blackwell handbook of personnel selection*, Blackwell Publishing Ltd, pp. 1-24 https://doi.org/10.1002/9781405164221.ch1.

Anderson, N, Herriot, P & Hodgkinson, GP 2001, 'The practitioner-researcher divide in Industrial, Work and Organizational (IWO) psychology: Where are we now, and where do we go from here?', *Journal of Occupational & Organizational Psychology*, vol. 74, no. 4, pp. 391-411 https://doi.org/10.1348/096317901167451.

Argyris, C & Schön, D 1996, *Organizational learning II: Theory, method and practice*, Addison-Wesley.

Beer, M 2001, 'Why management research findings are unimplementable: An action science perspective', *Reflections*, vol. 2, no. 3, pp. 58-65.

Bennis, WG & O'Toole, J 2005, 'How business schools lost their way', *Harvard Business Review*, vol. 83, no. 5, pp. 96-104 https://hbr.org/2005/05/how-business-schools-lost-their-way.

Bhaskar, R 2013, *A realist theory of science*, Routledge.

Buckley, RM., Norris, AC & Wiese, DS 2000, 'A brief history of the selection interview: may the next 100 years be more fruitful'. *Journal of Management History (Archive)*, vol. 6, no. 3, pp. 113-126 https://doi.org/10.1108/EUM0000000005329.

Cascio, WF & Aguinis, H. 2008, 'Research in industrial and organizational psychology from 1963 to 2007: Changes, choices, and trends', *Journal of Applied Psychology*, vol. 93, no. 5, pp. 1062-1081 https://doi.org/10.1037/0021-9010.93.5.1062.

Chalmers, AF 2013, *What is this thing called science?* Hackett Publishing.

Creswell, JW & Miller, GA 1997, 'Research methodologies and the doctoral process', *New Directions for Higher Education*, vol. 1997, no. 99, pp. 33-46 https://doi.org/10.1002/he.9903.

Dana, J, Dawes, R & Peterson, N 2013, 'Belief in the unstructured interview: The persistence of an illusion', *Judgment and Decision*

Making, vol. 8, no. 5, pp. 512-520 https://journal.sjdm.org/12/121130a/jdm121130a.pdf.

Delanty, D & Strydom, P 2003, 'Introduction: What is the philosophy of social science?' in D Delanty & P Strydom (eds.), *Philosophies of social science: The classic and contemporary readings*, Open University Press, pp. 1-12.

Diab, DL, Pui, SY, Yankelevich, M & Highhouse, S 2011, 'Lay perceptions of selection decision aids in US and non-US samples', *International Journal of Selection and Assessment*, vol. 19, no. 2, pp. 209-216 https://doi.org/10.1111/j.1468-2389.2011.00548.x.

Dipboye, RL 2007, 'Eight outrageous statements about HR science', *Human Resource Management Review*, vol. 17, no.2, pp. 96-106 https://doi.org/10.1016/j.hrmr.2007.04.001.

do Amaral, J. A. A. and Okazaki E. (2016) University students' support to an NGO that helps children with cancer: Lessons learned in thirteen academic projects. *International Journal of Action Research*, vol. 12 no. 1, p. 50.

Eden, C & Huxham, C 1996, 'Action research for management research', *British Journal of Management*, vol. 7, no. 1, pp. 75-86. https://doi.org/10.1111/j.1467-8551.1996.tb00107.x.

Ghoshal, S 2005, 'Bad management theories are destroying good management practices', *Academy of Management Learning & Education*, vol. 4, no. 1, pp. 75-91 https://doi.org/10.5465/amle.2005.16132558.

Gill, MJ 2014, 'The possibilities of phenomenology for organizational research', *Organizational Research Methods*, vol. 17. no. 2, pp. 118-137 https://doi.org/10.1177/1094428113518348.

Greenwood, DJ & Levin, M 2007, *Introduction to action research: Social research for social change*, (2nd edn), Sage Publications, Thousand Oaks https://dx.doi.org/10.4135/9781412984614.

Guion, RM 1998, 'Some virtues of dissatisfaction in the science and practice of personnel selection', *Human Resource Management Review*, vol. 8, no. 4, pp. 351-365 https://doi.org/10.1016/S1053-4822(99)00004-2.

Guzman, GAC & Wilson, J 2005, 'The "soft" dimension of organizational knowledge transfer', *Journal of Knowledge Management*, vol. 9, no. 2, pp. 59-74 https://doi.org/10.1108/13673270510590227.

Hambrick, DC 1994, 'What if the Academy actually mattered?' *Academy of Management Review*, vol. 19, no. 1, pp. 11-16 https://doi.org/10.5465/amr.1994.9410122006.

Harley, B 2015, 'The one best way? 'Scientific' research on HRM and the threat to critical scholarship', *Human Resource Management Journal*, vol. 25, no. 4, pp. 399-407 https://doi.org/10.1111/1748-8583.12082.

Herriot, P 1992, 'Selection: The two subcultures', *European Work and Organizational Psychologist*, vol. 2, no. 2, pp. 129-140 https://doi.org/10.1080/09602009208408539.

Herriot, P 1993, 'A paradigm bursting at the seams', *Journal of Organizational Behavior*, vol 14, no. 4, pp. 371-375 https://www.jstor.org/stable/2488288.

Highhouse, S 2008a 'Facts are stubborn things', *Industrial and Organizational Psychology*, vol. 1, no. 3, pp. 373-376 https://doi.org/10.1111/j.1754-9434.2008.00069.x.

Highhouse, S 2008b, 'Stubborn reliance on intuition and subjectivity in employee selection', *Industrial and Organizational Psychology*, vol. 1, no. 3, pp. 333-342 https://doi.org/10.1111/j.1754-9434.2008.00058.x.

Highhouse, S, Brooks, ME, Nesnidol, S & Sim, S 2017, 'Is a .51 validity coefficient good? Value sensitivity for interview validity', *International Journal of Selection and Assessment*, vol. 25, no 4, pp. 383-389 https://doi.org/10.1111/ijsa.12192.

Hodgkinson, GP 2011, 'Why evidence-based practice in I-O psychology is not there yet: Going beyond systematic reviews', *Industrial and Organizational Psychology*, vol. 4, no. 1, pp. 49-53 https://doi.org/10.1111/j.1754-9434.2010.01294.x.

Hodgkinson, GP & Starkey, K 2011, 'Not simply returning to the same answer over and over again: Reframing relevance', *British Journal of Management*, vol. 22, no. 3, pp. 355-369 **https://doi.org/10.1111/j.1467-8551.2011.00757.x**

Hoffman, AJ 2016, 'Reflections: Academia's emerging crisis of relevance and the consequent role of the engaged scholar', *Journal of Change Management*, vol. 16, no. 2, pp. 77-96 https://doi.org/10.1080/14697017.2015.1128168.

Huffcutt, AI & Arthur, W 1994, 'Hunter and Hunter (1984) revisited: Interview validity for entry-level job', *Journal of Applied Psychology*, vol. 79, no. 2, pp. 184-190 https://doi.org/10.1037/0021-9010.79.2.184.

Huffcutt, AI & Culbertson, SS 2011, 'Interviews', in S. Zedeck (ed.), *APA handbook of industrial and organizational psychology*, Vol 2: Selecting and Developing Members for the Organization, American Psychological Association, pp. 185-203.

Huffcutt, AI, Culbertson, SS & Weyhrauch, WS 2014, 'Moving forward indirectly: Reanalyzing the validity of employment interviews with indirect range restriction methodology', *International Journal of Selection and Assessment*, vol. 22, no. 3, pp. 297-309 https://doi.org/10.1111/ijsa.12078.

Hughes, T, Bence, D, Grisoni, L, O'regan, N & Wornham, D 2011, 'Scholarship that matters: Academic-practitioner engagement in business and management'. *Academy of Management Learning & Education*, vol. 10, no. 1, pp. 40-57 https://doi.org/10.5465/amle.10.1.zqr40.

Husserl, E 1973, *Experience and judgement: Investigations in a genealogy of logic* (J. S. Churchill & K. Ameriks, Trans.), Northwestern University Press.

König, C.J, Klehe, UC, Berchtold, M & Kleinmann, M 2010, 'Reasons for being selective when choosing personnel selection procedures', *International Journal of Selection and Assessment*, vol. 18, no. 1, pp. 17-27 https://doi.org/10.1111/j.1468-2389.2010.00485.x.

Kuncel, NR 2008, 'Some new (and old) suggestions for improving personnel selection', *Industrial and Organizational Psychology*, vol. 1, no. 3, pp. 343-346 https://doi.org/10.1111/j.1754-9434.2008.00059.x.

Lincoln, YS, Lynham, SA & Guba, EG 2011, 'Paradigmatic controversies, contradictions, and emerging confluences, revisited', in Denzin NK & Lincoln YS (eds.), *The SAGE handbook of qualitative research*, vol. 4, pp. 97-128), Sage Publications, Inc.

McKay, J & Marshall, P 2001, 'The dual imperatives of action research', *Information Technology & People*, vol. 14, no. 1, pp. 46-59 https://doi.org/10.1108/09593840110384771.

Meacham, JA 1983, 'Wisdom and the context of knowledge: Knowing that one doesn't know', in D Kuhn & JA Meacham (eds.), *On the development of developmental psychology*, Vol. 8, Contrib Hum Dev. Basel, Karger, pp. 111-134 https://doi.org/10.1159/000407340.

Mintzberg, H & Gosling, J 2002, 'Educating managers beyond borders', *Academy of Management Learning & Education*, vol. 1, no. 1, pp. 64-76 https://doi.org/10.5465/amle.2002.7373654.

Moustakas, C 1994, *Phenomenological research methods*, Sage Publications, Inc.

Muchinsky, PM 2004, 'When the psychometrics of test development meets organizational realities: A conceptual framework for organizational change, examples, and recommendations', *Personnel Psychology*, vol. 57, no. 1, pp. 175-209 https://doi.org/10.1111/j.1744-6570.2004.tb02488.x.

Nowotny, H, Scott, P & Gibbons, M 2001, *Re-thinking science. Knowledge and the public in an age of uncertainty*, Polity Press www.scielo.org.ar/pdf/cts/v1n1/v1n01a14.pdf.

Pettigrew, AM 1997, 'The double hurdles for management research', in T Clark (ed.), *Advancement in organizational behaviour: Essays in honour of Derek S Pugh* Dartmouth Press, pp. 277-296.

Pettigrew, AM 2001, 'Management research after modernism', *British Journal of Management*, vol. 12, no. S1), pp. S61-S70 https://doi.org/10.1111/1467-8551.12.s1.8.

Phillips, JM & Gully, SM 2008, 'The role of perceptions versus reality in managers' choice of selection decision aids', *Industrial and Organizational Psychology*, vol. 1, no. 3, pp. 361-363 https://doi.org/10.1111/j.1754-9434.2008.00064.x.

Rauschenberger, JM & Schmidt, FL 1987, 'Measuring the economic impact of human resource programs', *Journal of Business and Psychology*, vol. 2, pp. 50-59 https://doi.org/10.1007/BF01061524.

Romme, AGL, Avenier, MJ, Denyer, D, Hodgkinson, GP, Pandza, K, Starkey, K & Worren, N 2015, 'Towards common ground and trading zones in management research and practice', *British Journal of Management*, vol. 26, no. 3, pp. 544-559 https://doi.org/10.1111/1467-8551.12110.

Rousseau, DM & Barends, EGR 2011, 'Becoming an evidence-based HR practitioner', *Human Resource Management Journal*, vol. 21, no. 3, pp. 221-235 https://doi.org/10.1111/j.1748-8583.2011.00173.x.

Ryan, AM, McFarland, L & Baron, H 1999, 'An international look at selection practices: Nation and culture as explanations for variability in practice', *Personnel Psychology*, vol. 52, no. 2, pp. 359-392 https://doi.org/10.1111/j.1744-6570.1999.tb00165.x.

Rynes, SL 2007, 'Editor's afterword: Let's create a tipping point: What academics and practitioners can do, alone and together', *Academy of Management Journal*, vol. 50, no. 5, pp. 1046-1054 https://doi.org/10.5465/amj.2007.27156169.

Rynes, SL 2012, 'The research-practice gap in I/O psychology and related fields: Challenges and potential solutions', in S. WJ Koslowski (ed.), *The Oxford handbook of organizational psychology*, Oxford University Press, Vol. 1, pp. 409-452.

Rynes, SL, Bartunek, JM, & Daft, RL 2001, 'Across the great divide: Knowledge creation and transfer between practitioners and academics', *Academy of Management Journal*, vol. 44, no.2), pp. 340-355 https://doi.org/10.5465/3069460.

Rynes SL, Colbert, AE., & Brown, KG 2002, 'HR professionals' beliefs about effective human resource practices: Correspondence between research and practice', *Human Resource Management*, vol. 41, no. 2, pp. 149-174 https://doi.org/10.1002/hrm.10029.

Rynes, SL, Giluk, TL, & Brown, KG 2007, 'The very separate worlds of academic and practitioner periodicals in human resource management: Implications for evidence-based management', *The Academy of Management Journal*, vol. 50, no. 5, pp. 987-1008 https://doi.org/10.5465/amj.2007.27151939.

Sanders, K, van Riemsdijk, M, & Groen, B 2008, 'The gap between research and practice: A replication study on the HR professionals' beliefs about effective human resource practices', *The International Journal of Human Resource Management*, vol. 19, no. 10, pp. 1976-1988 https://doi.org/10.1080/09585190802324304.

Sayer, AR 2010, *Method in social science: A realist approach* (2nd ed.), Routledge, London https://doi.org/10.4324/9780203850374.

Shackleton, V & Newell, S 1994, 'European management selection methods: A comparison of five countries', *International Journal of Selection and Assessment*, vol. 2, no. 2, pp. 91-102 https://doi.org/10.1111/j.1468-2389.1994.tb00155.x.

Shapiro, DL & Kirkman, B 2018, 'It's time to make business school research more relevant', *Harvard Business Review*, July 19,viewed November 30, 2020 from https://hbr.org/2018/07/its-time-to-make-business-school-research-more-relevant.

Skarlicki, DP, Latham, GP, & Whyte, G 1996, 'Utility analysis: Its evolution and tenuous role in human resource management decision making', *Canadian Journal of Administrative Sciences/Revue Canadienne des Sciences de l'Administration*, vol. 13, no. 1, pp. 13-21 https://doi.org/10.1111/j.1936-4490.1996.tb00098.x.

Starbuck, WH 2006, *The production of knowledge: The challenge of social science research*, Oxford University Press https://doi.org/10.1093/acprof:oso/9780199288533.001.0001.

Starkey, K, & Madan, P 2001, 'Bridging the relevance gap: Aligning stakeholders in the future of management research', *British Journal of Management*, vol. 12, no. Special Issue), S3-S26 https://doi.org/10.1111/1467-8551.12.s1.2.

Thagard, P & Shelley, C 1997, 'Abductive reasoning: Logic, visual thinking, and coherence', in MLD Chiara, K Doets, D Mundici, & J v Benthem (eds.), *Logic and scientific methods*,Synthese Library vol. 259,Springer Science and Business Media, Dordrecht, pp. 413-427, Kluwer https://doi.org/10.1007/978-94-017-0487-8_22.

Tranfield, D, Denyer, D, & Smart, P 2003, 'Towards a methodology for developing evidence-informed management knowledge by means of systematic review', *British Journal of Management*, vol. 14, no. 3, pp. 207-222 https://doi.org/10.1111/1467-8551.00375.

Tranfield, D, & Starkey, K 1998, 'The nature, social organization and promotion of management research: Towards policy', *British Journal of Management*, vol. 9, no. 4, pp. 341-353 https://doi.org/10.1111/1467-8551.00103.

Van de Ven, AH & Johnson, PE 2006, 'Knowledge for theory and practice', *Academy of Management Review*, vol. 31, no. 4, pp. 802-821 https://doi.org/10.5465/amr.2006.22527385.

Weick, KE 2005, 'Organizing and failures of imagination', *International Public Management Journal*, vol. 8, no. 3, pp. 425-438 https://doi.org/10.1080/10967490500439883.

Weiss, HM & Rupp, DE 2011, 'Experiencing work: An essay on a person-centric work psychology', *Industrial and Organizational Psychology*, vol. 4, no. 1, pp. 83-97 https://doi.org/10.1111/j.1754-9434.2010.01302.x.

Wonderlic, EF 1937, Personnel as a control function, *Personnel*, vol. 14 no. 1, pp. 31-41.

Yates, FJ, Veinott, ES & Patalano, AL 2003, 'Hard decisions, bad decisions: On decision quality and decision aiding', in SL Schneider & J Shanteau (eds.), *Emerging perspectives on judgment and decision research* Cambridge University Press, pp. 13-63 https://doi.org/10.1017/CBO9780511609978.003.

Zhang, DC, Highhouse, S, Brooks, ME & Zhang, Y 2018, 'Communicating the validity of structured job interviews with graphical visual aids',

International Journal of Selection and Assessment, vol. 26, no. 2-4, pp. 93-108 https://doi.org/10.1111/ijsa.12220.

Zibarras, LD & Woods, SA 2010, 'A survey of UK selection practices across different organization sizes and industry sectors', *Journal of Occupational and Organizational Psychology*, vol. 83, no. 2, pp. 499-511 https://doi.org/10.1348/096317909X425203.

Biography

Edward "Ed" Hyatt has held multiple academic and professional roles over the course of his working life and is currently a senior research fellow at George Mason University. He has over a decade of research experience and more than seven years of managerial experience in the public procurement profession working for U.S. government agencies and universities. He holds several advanced degrees including a PhD in Business Management with a focus on Human Resource Management from The University of Melbourne. His research interests are centered on topics germane to organizational behavior and human resources such as personnel selection and managerial decision-making, and he also has a passion for research methods.

Researching resilience in action: Matters of action research as 'matters of care'

Akina Mikami

Abstract

This paper traces my action research (AR) journey and suggests a care perspective to AR. In my doctoral research that investigated the idea of resilience in disaster recovery, I engaged in a five-year collaborative inquiry of what it means to do resilience with a civil society group in Cairns, Australia, which acts alongside Fukushima children affected by the 2011 nuclear disaster. AR enabled co-generative and democratizing ways of knowledge creation in which research can be simultaneously action-driven, theory-advancing and policy-informing. In my project, I addressed matters of action and inquiry as 'matters of care' – a generative engagement with neglected everyday needs and its wider entanglements that are tied to a multigenerational problem. Through this lens, I trace how I came to do AR, encountered a problem, found my co-researchers and engaged in collaborative inquiry. Finally, I reflect on who this research is for, before concluding with 'What's next?'

Key words: Action research PhD, action research thesis, AR research journey, matters of care

What is known about the topic?

There have been discussions about AR conducted in university setting (Greenwood and Levin, 2007: 251; Herr and Anderson, 2014), AR principles applied to PhD thesis (Dick, 1997) and some AR dissertation experiences (Klocker, 2012; Gibbon, 2002; Southby, 2017; Macguire, 1993; Ogawa, 2009). In conventional academic research setting, AR PhD is still seen as "the new kid on the block" (Herr and Anderson, 2014: 1) and more accounts of AR doctoral research experiences are considered beneficial. While care perspective to problems of social change and ecological sustainability has been discussed (Moriggi et al., 2020; Middleton and Samanani, 2021; Tozzi, 2021), a care perspective to Action Research and its application to doctoral research is less known.

What does this paper add?

This paper provides a reflective account of adopting an AR approach to doctoral research. It also offers a care perspective as one useful way to addressing a matter of inquiry and action.

Who will benefit from its content?

- PhD students potentially or currently doing AR as part of their research project.
- PhD students potentially or currently doing AR as part of their research project who are on part-time candidature and/or raising a family.
- Academic staff in university potentially or currently supervising an AR PhD project.

What is the relevance to AL and AR scholars and practitioners?

- This paper documents how a PhD student might adopt AR principles to various stages of doctoral research from encountering a research problem, initiating and conducting collaborative inquiry with research participants and producing a dissertation.
- This paper considers the ways that AR PhD research can contribute to action-taking, policy-informing and theory-building processes.
- This paper demonstrates that while unexpected encounters and blockages can be expected in AR PhD, they can also lead to new questions and breakthroughs.
- This paper discusses the usefulness of applying a care perspective to an AR problem/question.

Received   January 2022   Reviewed   June 2022   Published November 2022

Introduction

> All that I learned and felt in Cairns do not end here. When I go back to Fukushima, I want to continue projecting my "own voice".
> – Hikari (pseudonym)[1], 2017 participant, camp diary.

In the opening, I acknowledge the Wotjobaluk, Jaadwa, Jadawadjali, Jupagulk and Wergaia peoples, the Traditional Custodians of the lands on which I live, and the Mirarr people, the Traditional Custodians of the lands that became implicated in the Fukushima Daiichi Nuclear Power Plant accident, and pay my respects to their Elders past, present and emerging. I also pay my respect to those affected by the 2011 Tōhoku earthquake, tsunami and the subsequent nuclear accident and honour those who continue to rebuild their lives.

When I began my action research (AR) journey as a PhD project, my map was like a blank piece of paper with two dots on it: 'departure' and 'arrival.' With some key questions and some key AR concepts (Greenwood & Levin 2007; Ogawa 2009; Wadsworth 2014), just how, where, when and what kind of *lines* would be drawn to connect those dots was largely dependent on my collaborative inquiry with my co-researchers. Due to this relational and open-ended nature of AR, there were certainly unexpected moments and blockages. However, each challenge led to breakthroughs and next steps. I came to appreciate that it is the journey, more than the destination itself, that makes AR challenging and rewarding.

The 'why' question, however, was there from the beginning and grew in transnational and intergenerational scope as the journey unfolded. In the beginning, my 'why' was simple: a place where children can be outdoors without concerns for disaster-induced radiation risks. This specific everyday need emerged following the

---

1 All individuals' names are pseudonyms except for people in public offices and also Maki, which is a real name used with permission. Smile with Kids (SWK) is also a real name, which is used with permission.

2011 nuclear disaster in Japan. I saw ordinary people just like me, many of them juggling care and work responsibilities, act and reflect on what can be done to protect and sustain a sense of normality in the lives of children. Yet, as I will discover, the problem was not really a matter of organizing short-term trips to minimize radiation exposure. Responding to this specific need required a generative engagement to 'stay with trouble' (Haraway 2016) with the wider entanglements that shaped that need. This included engaging with the highly contested ideas of resilience in disaster recovery politics: What can be done to stay resilient in the *long durée* of nuclear emergency? What does it mean to 'build back better' from a nuclear disaster? What happens if the means to stay resilient becomes undermined by the very practices done in the name of resilience? How can resilient futures be reimagined?

Starting with a response to this everyday need, I took the first step, and then the next step, and kept going as I *felt* my way through. What transpired was a dynamic and transformative journey of collaborative inquiry experiences across the next five years of my life, which has significantly shifted my view of what research can do. My background is in the interdisciplinary field of media and communication studies that builds on the disciplines of anthropology, sociology and human geography. Rooted in critical paradigm, I am trained to investigate social, cultural and political life across spaces and places from the lens of power, inequality and social change. Doing AR enabled me to engage in co-generative and democratizing ways of knowledge creation, in which research can be simultaneously action-driven, theory-advancing and policy-informing. I have been especially inspired by the ideas of research process as democratization (Greenwood & Levin 2007) and a commitment to the flourishing of persons, communities and the wider ecology (Bradbury 2015, p. 12). In addition, because AR entails a dynamic form of inquiry that engages with real people on a real problem, I found it helpful to approach inquiry as 'living systems' (Wadsworth 2014). Furthermore, I have been challenged by the idea of research as co-creating 'action narratives' that produce 'civic knowledge' (Ogawa 2020, pp. 6-7). As my project examined how grassroots civil society practice informs the policy-oriented notion of resilience, Ogawa's idea of action narratives that challenge dominant policy discourse from citizens' perspectives became important.

My doctoral research examined how civil society practice shapes resilience in disaster recovery. From 2017 to 2021, employing a case study methodology, I engaged in a collaborative inquiry of what it means to *do* resilience with a civil society group in Cairns, Australia. This group acts alongside Fukushima children affected by the Tōhoku earthquake, tsunami and the subsequent Tokyo Electric Power Company's (TEPCO) nuclear power plant accident on March 11, 2011 (hereafter '3.11'). Applying an AR approach to this project not only led to important insights on how civil society shapes alternative ideas of resilience. It also positioned me to become a co-subject in the real-time making of resilient futures that centered on the hopes and concerns voiced by Fukushima students. Beside my positionality as a Japanese woman living in Australia, raising a family and doing PhD part-time, the evolving researcher positionality throughout the journey illuminated an interlacing of a generational trajectory.

In this paper, I trace my AR journey and offer a contribution of care perspective to AR. First, I discuss the idea of matters of AR as 'matters of care'. Next, through this lens, I trace how I came to do AR, encountered a problem, found my co-researchers and engaged in collaborative inquiry, which then materialized into a thesis. Finally, I reflect on who this research is for before concluding with 'What's next?' I hope that my story will encourage new students to embark upon their own AR journeys.

Matters of AR as 'matters of care'

One of the privileges of doing AR is not only applying AR principles to a project but also contributing to the AR field. In my project, I encountered an approach to matters of action and inquiry as 'matters of care' (Puig de la Bellacasa 2017). I suggest this perspective can be helpful to other AR projects. I especially see its usefulness in tackling a problem that is presented as a specific unmet need in an everyday context, which cannot be readily solved by more technoscientific solutions, more 'best practices,' or more deconstructionist critiques – as important as they all are. A care perspective involves a generative engagement with a specific need, and the wider ethico-political entanglements that produce that need, in the situated experiences of everyday living. Recently, I noticed that care perspective is attracting attention as one fruitful way to research social and environmental problems of change and sustainability, showing relevance of this perspective to AR (Moriggi et al. 2020; Tozzi 2021; Middleton & Samanani 2021). By care, I refer to:

> …a species activity that includes everything that we do to maintain, continue, and repair our "world" so that we can live in it as well as possible. That world includes our bodies, our selves, and our environment, all of which we seek to interweave in a complex, life-sustaining web (Fisher & Tronto, 1990, p. 40; Tronto, 1993, p. 103).

Like AR, care is an action that is committed to human flourishing and environmental sustainability and is premised on relational mode of shaping the world. In particular, María Puig de la Bellacasa (2017) proposes a care perspective to reimagine research from the viewpoints of post-humanist feminism and Science and Technology Studies (STS). Building on Collier and Lakoff's (2005) notion of 'regimes of living,' she asks how research can better facilitate 'reflection and action in situations in which "living" has been rendered problematic' (Collier & Lakoff 2005, p. 22). In today's shifting ontologies of risks, the question of how to better respond to socionatural problems such as nuclear disaster, pollution, pandemic or climate change necessarily raises ethical problems on what it means to live well and shape better futures. To address this question, Puig de la Bellacasa (2017) draws on Bruno Latour's (2004) work on 'matters of facts' and 'matters of concern' to suggest a 'matters of care' perspective (Puig de la Bellacasa 2017; Latour 2004). 'Matters of facts' are staging of reality as self-evident, ahistorical and indisputable, often presented as diagrams or percentage numbers that can be 'proven' by positivist scientific techniques. In contrast, 'matters of concern' is the staging of reality as contested, tentative and politicized as various actors gather around the 'matters of facts' to raise questions, worries and concerns around the problem. In response, Puig de la Bellacasa (2017) proposes 'matters of care' as the staging of reality as *alive* and *living*. It indicates a situated and committed form of inquiry that seeks to sustain the everyday world while also opening it up to new possibilities and political stakes. As Middleton and Samanani (2021) assert, addressing a problem through a care lens calls for:

> …close and sustained attention to the ongoing production of meaning, and of the possibilities for acting alongside others, within particular situated contexts. Simultaneously, it demands that we look beyond this everyday terrain, to trace the wider entanglements that shape it, and to hold these entanglements in generative tension with everyday commitments (p. 33).

My project responded to a need for a place where children can be outdoors free from disaster-induced radiation concerns following the nuclear disaster. This problem could not be adequately addressed by more 'matters of facts,' which sought to solve the problem by presenting that Fukushima beaches are now 'safe' because the radiation air doses are measuring below 0.1 microsieverts an hour. This problem could also not be fully addressed by more 'matters of concern,' which raised questions about safety standards and its link with the government agendas to revitalize the economy of a disaster-affected region and restart nuclear energy. This problem required a 'matters of care' response, which asked what can be done in the everyday context so children can be outdoors without radiation concerns while also being attentive to the wider forces that reproduce this need.

During the course of my journey, many lines were drawn on my AR map. Looking back, I see a single line of inquiry that underscored my entire AR journey in the form of these two questions: How can research center the hopes and concerns voiced by Fukushima children in the lived experiences of disaster recovery politics, trace the wider problems that shape those concerns, and hold these tensions in critical hope in the ongoing everyday commitments? How can research engage in the real-life actions that value their voices, not only to gain insight, but to partake in the reimagining of resilient futures?

In the following, I trace my AR journey through the lens of care by beginning with an account of how I came to do AR project. It began from an unexpected encounter when I was looking for ways to value the voices of Fukushima people in disaster resilience process.

How I came to do an AR PhD

My AR journey began with an unexpected encounter with Professor Akihiro Ogawa (hereafter Ogawa sensei). In March 2016, on the fifth anniversary of 3.11, his name came up when I was corresponding with a Fukushima-based civil society organization (CSO) about a Japanese-English translation work. Until then, I was acting and researching – but not together. I strongly believed that the voices of Fukushima people, especially women and children, were crucial to rebuilding post-3.11 futures. To do this, I was 'acting' by doing volunteer translations for a CSO that amplified their voices. Separately, I was 'researching' by doing a PhD project on CSOs' communication practices in the context of post-3.11 with another supervisor in another department. I had considered AR. However, when I suggested to my then supervisor the idea of Participatory Action Research – something I learnt in my previous job doing monitoring and evaluation of Communication for Development initiatives, I was advised that such an approach is more practitioner-oriented. To be 'more academically rigorous,' I was encouraged instead to conduct expert interviews and focus on theoretical development. As I wanted to do theory-building *and* action-taking, I had kept them separate.

This was why I was so surprised to hear about Ogawa sensei. How did I not know that in the same university there was someone specializing in civil society and AR with an interest in post-3.11 Japan? It turns out he had moved to Melbourne six months prior and was based in another institute. I immediately arranged a meeting. He was approachable, open-minded and had shared concerns about the situations unfolding in Japan. He showed me the book, *Introduction to Action Research: Social Research for Social Change* by Davydd Greenwood and Morten Levin (Greenwood & Levin 2007). Talking with him, I started to wonder: What if there is a possibility for me to do both action *and* research? At that time, in my candidature, I was at the stage of post-Confirmation and pre-data collection. I was about to go on maternity leave with a view to start fieldwork the following year. Although I did not take supervision change lightly, I felt I was at a critical turning point. The following month, I took my first step and asked Ogawa sensei to become my supervisor. Upon reading my research proposal to do AR of CSO's communication practices, he agreed.

In August, to learn more about AR, I participated in the AR subject he coordinated. As I was still on maternity leave, I enrolled informally as a visiting student. Based in a regional town, commuting to the university campus took five hours one way on bus, train and tram. Leaving home before sunrise and returning back home close to midnight, attending these sessions with a two-month-old was a commitment. Sleepy, tired yet excited, I went to these sessions and actively participated. Exposed to the rich theoretical foundations of AR and a practical application of AR in a teamwork exercise, I became excited about the possibilities.

Encountering a matter of action and inquiry

I took my next significant step when I decided to respond to a specific need that persisted in the everyday context of 3.11 disaster recovery. I was sitting outside on a cool spring morning watching my two-year old gleefully jump in muddy puddles. My three-month old had just settled to sleep, and I was using the little window of opportunity to scroll through an e-newsletter of a CSO in Japan. I came across photos and stories about 'recuperation' activities. This was a grassroots endeavour that created opportunities for children living in areas with disaster-induced radiation concerns to spend weekends and school holidays in another location with the aim to refresh their bodies and minds. Inspired by the state-run recuperation activities being held in

Belarus and Ukraine after the 1986 Chernobyl accident, civil society groups began trialling similar activities. There were photos of children splashing in the sea, pulling out daikon radish from the soil, eating freshly pounded *mochi* (rice cake), feeling worms on their little hands and making a snowman in the snow. I felt stirred to know more. At the time, I was still planning on researching how citizen voice is shaped by CSOs' communication practices. However, I kept being drawn to recuperation practice and finding myself asking: What can I do? Are there recuperation activities in Australia?

I shifted my research focus. I decided to investigate how resilience is enacted by recuperation practice in the 3.11 disaster recovery context. First, I was struck by how difficult the situation seemed to sustain these activities. That year, an umbrella CSO that coordinates a network of recuperation groups released a report on the state of recuperation activities in Japan based on nationwide surveys (3.11 Japan Nuclear Disaster Aid Association, 2016). I found out that despite an ongoing demand, the majority of recuperation groups were growing extremely weary because they were run by unpaid volunteers with limited resources, time and funds. I was shocked to learn that although there was a state-sponsored program supporting outdoor experiences for Fukushima children, hardly any groups were eligible to apply. To access the state subsidies, the camps had to be held either *within* Fukushima or have duration longer than a week if held outside Fukushima. These were near impossible requirements for many recuperation groups, which held activities outside Fukushima and were organized by volunteers who could hardly take weeklong leaves from care and work commitments. Furthermore, I was troubled by social stigma and political sensitivities around these activities. I read about parents signing up their children to camps in secret to avoid being perceived as 'anti-science' or 'anti-reconstruction'. I also read about recuperation practices receiving criticisms because *either* it was seen as fanning fear that radiation risk is dangerous when no one died from it *or* short-term camps were considered inadequate to reduce radiation exposure. Meanwhile, I discovered that the demand for recuperation activities might possibly grow. The Japanese government had announced to cut housing subsidies for evacuees from March 2017 onwards, forcing many evacuees to return to the affected region.

I also felt unsettled by how little literature there was about recuperation practice. Apart from some grey literature, a few journal articles and personal testimonies, hardly anything was written about it. This was in stark contrast to many other civil society activities that were receiving attention in mainstream disaster resilience policies and academic research at the time. In the previous year, Japan had just hosted the United Nations World Conference on Disaster Risk Reduction, which led to the establishment of the 2015-2030 Sendai Framework, the global blueprint of disaster risk reduction. Amidst the vibrant discussions on the roles that civil society plays in resilience-building in disaster recovery, there was deafening silence around recuperation practice.

To contribute to policymaking and theory development, I decided to frame my research around the notion of resilience. I strongly felt that sustaining the option to participate in recuperation practice was about recognizing the self-determination of people who chose to live in Fukushima while also choosing to *not* accept disaster-induced radiation as part of their everyday life. Therefore, recuperation practice provided one important means to stay resilient in the lasting impacts of nuclear disaster. However, the voices of recuperation participants and organizers were clearly not reflected in the disaster resilience policies (Hikita 2018, pp. 114–124). Furthermore, in the social sciences, theoretical debates on resilience as a form of neoliberal governmentality were on the rise (Chandler & Reid 2016). Tracing the change from a 'problem-focused' vulnerability paradigm in the 1990s to a 'capacity-oriented' resilience approach from 2000s onwards, these debates alerted that the seemingly benevolent shift in focus from what prevents people from withstanding disasters to what people can do to recover for themselves signalled new forms of domination and exclusion. The emerging discussions criticized knowledge construction that confounded human agency with market participation and warned the formation of resilient subjectivity based on adaptive capacity of individuals that mystified systemic inequalities as self-help opportunities. While such critical stance was warranted, I wanted to move past the neoliberal impasse to rethink how resilience can be reimagined at the grassroots scale. I wanted to engage with resilience as not merely a concept but an *already existing action*. I felt that inquiring with a recuperation group on what it means to *do* resilience might contribute to this important direction.

Finding my co-researchers

I found my co-researchers unexpectedly when I was talking to a friend in Melbourne in early January 2017. She told me about her friend, Maki, who had been organising camps in Cairns for several junior high school students from Fukushima. At the time, I was facing a blockage with finding a recuperation group in Australia. I contacted the CSO in Japan that coordinates a recuperation groups network. Although the representative was encouraging of my research and emailed me some literature, they were not affiliated with any overseas group as it was beyond their scope. I found a CSO in Sydney, and they were also supportive of my project, but 2016 was their final year of activity. I heard in passing there was some activity in Canberra but I could not find any contacts.

I was therefore all the more grateful that a path opened up. I asked my friend for an introduction and immediately followed up with a phone call. Maki was friendly and sincere and told me about SWK (Smile with Kids), the charity association she founded with several Japanese women in Cairns that organizes 'refreshment camps' for junior high school students living in Fukushima, Japan. I mentioned that I was doing PhD research on civil society's role in disaster recovery with a focus on recuperation activities and expressed interest in getting involved as a volunteer. Maki asked what my interests and skills were. I replied that I was interested in taking action together and learning together and offered Japanese-English translations skills. Maki said more assistance with translations would be helpful especially leading up to the annual fundraiser event and the annual camp. Feeling glad that what I could offer matched what they needed, I became involved with SWK as a volunteer translator and researcher.

My first translation task was a message from Minami (pseudonym), one of the four students that participated in SWK's inaugural camp in 2015. This was one of many 'voices' that mattered to SWK, with which I came to engage as 'matters of care' in this AR journey. Since then, I translated hundreds of Fukushima children's speeches; their messages for the Cairns community for the annual fundraiser events; their daily diaries and exchanges with the local community in Cairns during the camps; and messages after they returned to Fukushima. I also translated many messages from participants' mothers expressing concerns and hopes about situations in Fukushima. Moreover, I translated from English to Japanese two video messages of Yvonne Margarula, a Senior Traditional Owner of the Mirarr people in the Kakadu region, as part of SWK's activity.

Translating Minami's message made me wonder about the camp. In this message, Minami had exciting news to share with the SWK community. She said she was so inspired by the experiences in Cairns that she decided to study abroad for the next ten months. She explained about hearing a story about high school students in Germany discussing the Fukushima nuclear accident from multiple perspectives and deliberating on why it could not be prevented. She mentioned she wanted to go overseas again and learn how "to think more independently and holistically" about what is happening around her. Moreover, she emphasized that the experience of doing a speech in English in front of people in Cairns changed her. She said that although she felt nervous at first, her host family encouraged her to be confident and helped her practise. She asserted: "It made me realize how difficult yet how important it is to have my voice heard". Translating her message gave me the impression that SWK's camp was more than just a short-term trip to reduce radiation exposure risks.

After a month of interacting with Maki and doing some Japanese-English translation tasks, I decided to initiate a collaborative inquiry with SWK. As my project examined the grassroots enactment of resilience, I became interested in how their practices enacted and contested the idea of resilience. Adopting an 'action-mindset' (Greenwood 2007; Ogawa 2009), I was getting involved to not merely understand but also actively participate as a co-subject in their action-reflection cycles (Greenwood & Levin 2007, p. 66). I wanted to find out more about who else make up the SWK community, how they relate with Fukushima children, how they view the idea of 'building back better' from 3.11 and what challenges and opportunities they were facing. Since recuperation practice primarily aims to secure a place where children can spend time outdoors free from disaster-induced radiation concerns, I wanted to explore how such a place was being created in SWK's activities. Moreover, a majority of recuperation activities were held in Japan for children aged around kindergarten to primary school years. I became interested in what is distinct about the site of Cairns, Australia, and the focus on junior high school students aged between 12 and 15. I expressed my interest to participate in the camp and Maki welcomed the idea.

Engaging in collaborative inquiry

From March 2017 to December 2021, I engaged in a collaborative inquiry with my co-researchers on how to do resilience in disaster recovery by focusing on what it means to create a place for children free from disaster-induced radiation concerns. I deployed various research methods. These included participant observation, interviews, diaries, collaborative material collection (news articles, photographs, social media posts, documentaries and policy documents) and collaborative content production mainly through Japanese to English translations of SWK contents. I made two field trips to Cairns in 2017 and 2018 to participate in the camps. In addition, I corresponded regularly with my co-researchers through emails, phone calls and group chats, utilizing applications such as Messenger, LINE, Zoom and a closed Facebook group page dedicated to SWK volunteers.

Learning how to inquire together

The most important aspect of collaborative inquiry was relationship-building and learning *how* to inquire with my co-researchers. I spent time getting to know my co-researchers as individuals and appreciating the diversity within the SWK community. At the same time, I engaged in ongoing researcher reflexivity as my co-researchers were also finding out who I was. I remember the flight from Melbourne to Cairns as I headed towards my first field trip. As the plane took off, I felt excited but also cautious about the various dynamics and possible tensions. First, I was wary of the fine line between disaster research and disaster capitalism. I did not want to do, or be associated with, research that profits from people's hardships. Prior to fieldwork, I heard from my peers about the problem of 'research fatigue' (Clark 2008) experienced by the people of Fukushima as many researchers rushed to collect data from or about the disaster-affected community. Although my research focused on the organizer of recuperation practice (SWK) and not the participants (Fukushima children), I did not want to contribute to this problem. Moreover, while my research was motivated by a sense of wanting to help, I did not want to be associated with a saviour complex fuelled by a sense of self-gratifying heroism. The politics and ethics of responsiveness to others' needs was incredibly important and became something my co-researchers and I reflected attentively and continuously throughout the project.

I was also mindful of my positionality as a Japanese woman living in Australia, raising a family and doing a PhD part-time. As SWK was driven mainly by Japanese women residing in Australia, I was an insider given my native language fluency and competence to read cultural cues. Yet I was also an outsider as I was not based in Cairns and not embedded in the social networks. I tried to see this ambiguous insider-outsider positionality as an opportunity in which I was able to have the cultural proximity that enabled shared understandings but also an appropriate distance that generated new questions. Moreover, my AR journey coincided with my personal journey of raising a family, bringing distinct dynamics to the collaborative inquiry experience. I went to my first field trip with my two-year-old in one hand, my six-month old strapped to me and my partner beside me. I was mindful of the various dynamics of bringing a toddler and an infant to fieldwork (Brown & Dreby 2013) and the nuanced implications of my identity as a 'mother-researcher-feminist-woman' to the research process (Frost & Holt 2014). However, I found that within the SWK community, there was respect towards diverse expressions of identities and family life. None of my children, my partner nor I was boxed into being a certain way either individually or as family. Furthermore, perhaps due to all SWK volunteers juggling work and/or family responsibilities, there was shared understanding for everyone to be involved within their ongoing everyday commitments. This inclusivity and flexible temporality enabled me to find a certain pace of collaborative inquiry that was amenable to the syncopated rhythms of my everyday life. This distinct temporality also fitted well with my part-time candidature and afforded me to engage in the action-reflection cycle with my co-researchers over a longer duration.

To engage in collaborative inquiry with SWK, I experimented with different ways of acting and reflecting. This meant confronting my own prejudices about what it means to *know* and *do* things. In an earlier stage, I had preconceived ideas about what an action researcher does. I had an image of a researcher standing in front of the co-researchers and leading research. However, in my project, my co-researchers were already engaged in meaningful research. The idea of a researcher showing co-researchers how to generate knowledge felt inappropriate. I decided instead to focus on what Wadsworth (2015) calls 'shared inquiry capabilities'. Wadsworth argued that as humans there are shared desire and capacity to inquire about the world around us but also differing preferences on how we take in, process and act on information. She explained that if we tend to routinely approach particular inquiry capabilities that we favour over others, these preferences may over time turn into an 'inquiry style' (Wadsworth 2015, p. 754). Rather than imposing my inquiry style upon my co-researchers, I decided to learn how my co-researchers were inquiring about their world.

This was not a passive choice to merely follow their inquiry style. This was an active choice to resist the assumption that the researcher knows better and to value other ways of knowing and doing. Instead of standing in front of my co-researchers, I chose to stand *beside* them to inquire *with* them about the things that matter to them. When I later encountered the matters of care perspective, I realized that this relational way of co-creating knowledge resonated with what Puig de la Bellacasa describes as 'thinking-with' (2017, p. 71). It indicates an embodied, situated and speculative practice of knowledge generation that unfolds relationally between subjects.

In this relational mode of inquiry, I noticed a parallel collaborative inquiry happening. As I inquired with my co-researchers on what it means to build resilience, I was simultaneously engaging with the ways that my co-researchers were inquiring with *their* 'co-researchers' (Fukushima students) to envision better resilient futures. This process reminded me of a line drawn with a double line pen or a parallel pen, which has two different colours or two parallel points on the nib. When you glide the pen along the paper, the unfolding line is underscored by another colour or another line. Drawn this way, the line becomes clearer, sharper, more defined and with more dimension.

I found that an awareness of emotions, sensibilities and tactility was crucial to my co-researchers. It required '*deeper* listening' to hear what has not been heard or said or has not yet had a chance of expression (Wadsworth 2014, p. 91, emphasis in original). This evokes what Puig de la Bellacasa (2017) calls 'touching vision' (2017, p. 112) that inflects ways of knowing and envisioning through emotions and tactility. It mattered to my co-researchers when Fukushima students commented about the saltiness of seawater or the strange feeling of being rocked by the waves. It also mattered when the students voiced their hope to be able to swim in the sea of Fukushima again. It also mattered when Fukushima students raised concerns about the Japanese government's decision to discharge the treated radioactive water into the sea of Fukushima for the next 40 years. Engaging in collaborative inquiry, it became apparent that the various concerns and hopes voiced by Fukushima students became a matter of care that led to next steps.

Action-reflection cycles

As I experimented with how to do collaborative inquiry, I simultaneously began engaging in the action-reflection process with my co-researchers. At the 2017 camp, I noticed that SWK was re-evaluating its practice. In 2015 and 2016, there was a sense of urgency to restore normality in the lives of Fukushima children. However, from 2017 onwards, as time passed and the state-led reconstruction progressed, 3.11 was beginning to fade from public memory. Yet what SWK was hearing from their Fukushima participants was that disaster recovery was far from over. This problem became the first phase of an action-reflection cycle (2017-2019) that I engaged with the co-researchers. The guiding question was: What does it mean to continue recuperation activities when disaster-induced radiation concerns were increasingly becoming forgotten?

Engaging in this action-reflection process required recognizing the different views, emotions and sensibilities. Among the various views considered within SWK, questions were raised about whether SWK should identify with anti-nuclear activism. There was a view that without openly acknowledging the wider politics of why recuperation activities had to begin in the first place and why sustaining such activities became more difficult, SWK's work might become no different from other holiday programs or English language study programs. SWK was a charity association funded by donations and supported by the local community. Clear explanation was needed on why it was still relevant to create opportunities for children to be free from nuclear disaster-induced problems. However, while SWK was committed to exploring with Fukushima children a future that did not include another nuclear accident, the focus was not on a nuclear-free future. Moreover, there was a sentiment that it was not SWK's place to speak about the situation in Fukushima. SWK continued to hear directly from the Fukushima students about what is happening there and what can be done together.

In the action-reflection process, I noticed a dynamic of responding to the ongoing need for a place free from radiation concerns, while simultaneously recognizing the wider entanglements that reproduce this need. There was emphasis on inquiring with the Fukushima students about the matters they cared about – what they felt about spending time outdoors, what they considered to be problems and what they envisioned as better futures. As SWK centered the voices of Fukushima students in their cycle of planning, acting, observing and reflecting, I realized that my role as a volunteer translator was more than simply conveying Fukushima children's words from Japanese to English. I saw myself becoming an integral part of action.

In tandem with doing translation work, I shared my reflections with my co-researchers. In 2018, I organized a video call with the SWK committee members to share and seek feedback on my PhD research. I highlighted the ways that grassroots civil society practice reconfigured a place for children free from radiation concerns into a place where children can reconnect with nature and community, give an account of 3.11 from their everyday perspectives and imagine alternative environmental futures (Mikami 2021). In addition, when the committee decided to revise the organizational charter, I had an opportunity to provide my comments. One suggestion I made was to add 'exploring sustainable possibilities for the future' as one of the organizational objectives. Later that year, this point was further discussed at the Annual General Meeting and then adopted in the revised version. Despite the challenges to sustain recuperation practice, I remained hopeful. I was also learning to better inquire with the co-researchers.

Then came March 11, 2020. On the ninth anniversary of 3.11, the Covid-19 outbreak was declared a pandemic by the World Health Organization. At the time, I was expecting to return in two months to my doctoral candidature from another period of maternity leave. I had been planning on participating in the camp that was to be held in July that year. The pandemic not only caused the camp and the field trip to be cancelled due to travel restrictions and shifting the research landscape, but also exacerbated the ongoing problem of fading memories of 3.11. The question of what it means to build resilience and sustain recuperation practice took on an entirely new dimension. This became the second phase of an action-reflection cycle (2020-2021). In this phase, the problem was amplified in the lead-up to the previously postponed 2020 Tokyo Olympics and the United Nations Climate Change Conference (COP26). Otherwise known as the Reconstruction Olympics, the hosting of the international sport event was intended by the Japanese government to showcase to the rest of the world that the reconstruction process was on track. Moreover, in timing with COP26, nuclear energy was increasingly promoted as eco-friendly and economically efficient energy that can realize a zero-carbon society. Despite these challenges, SWK adapted its activities while continuing to center the voices of Fukushima children. From May 2021 to December 2021, I participated in the monthly Zoom sessions organized by SWK to discuss with Fukushima students what matters to them, what they are feeling and thinking, and what can be done together. These discussions led to doing a collaborative video project in timing with the tenth anniversary of 3.11 in March 2021. This video voiced a different narrative of disaster resilience and post-3.11 futures.

Co-generating a thesis

One of the rewarding aspects of AR was continuing the collaborative inquiry at the stage of thesis writing. To do this, I invited my co-researchers to my completion seminar and received their feedback before finalizing the chapters. As my collaborative inquiry experience foregrounded a distinct way of listening out for others' voices, I titled the thesis, *Resilience in a different voice*. In contrast to existing disaster resilience research, my dissertation provided a different account of how ideas of resilience are revealed, resisted and reimagined by a grassroots civil society practice in disaster recovery politics. Based on my collaborative inquiry experiences, I proposed a rethinking of resilience that centered on the voices of Fukushima children. In addition, because my thesis challenged the dominant policy discourse of resilience based on 'action narrative' (Ogawa 2020), the title also honoured AR and the co-creation of knowledge shaped by citizen voice. Furthermore, this thesis title acknowledged the legacy of care ethics. In particular, it paid tribute to Carol Gilligan's *In a Different Voice*, which is recognized as the first care ethics literature (Gilligan 1982). In her seminal thesis, which Harvard University Press calls "a little book that started a revolution", Gilligan argued that moral reasoning based on relational ontology of care offered a different voice or alternative ethics and should not be seen as inferior to the liberal justice traditions on independence (1982). Likewise, my thesis argued that knowledge of resilience, generated relationally through a care lens, was different but equally legitimate knowledge.

Who is my 'audience'?

> What has happened cannot be undone. But if we think about our future, and the future of our children and their children, there are things we can change from today. I want to pass on the gift of a better future to the children of tomorrow.
> – Rin (pseudonym), 2020-2021 participant, English speech.

> We borrow the earth from future children.
> – Engraved in a rock in Urabandai, Fukushima, Japan (see Figure 1).

Ogawa sensei often asked: 'Who is your audience?' I revisited this question as my thesis materialized. At the start, the audiences of my research were recuperation organizers, policymakers and social scientists that mobilize the resilience discourse in practice, policy and academic research. However, doing collaborative inquiry brought different audiences into view. I thought of each Fukushima student I met and their respective dreams and hopes. I thought of the Mirarr people and their hope for their lands and their future generations. I thought of the sea, soil, plants and animals: living and deeply interconnected. I thought of the co-researchers, Maki and the SWK community, who showed me what a small group of committed citizens can do to create a sense of wonder and open up possibilities for children affected by nuclear disaster.

Figure 1: An image of a rock in Urabandai, Fukushima, Japan, featured in SWK's video launched in timing with the tenth anniversary of 3.11 in March 2021. (Source: SWK YouTube)

As the project evolved to trace the wider entanglements pertaining to recuperation practice, I began thinking about who the audience is in light of the historical contingency of this research. I especially had to confront the historical problems tied to the social imaginary of nuclear power and the idea of a resilient nation. These included, but were not limited to, war memories of nuclear bombs; nuclear imperialism and settler-colonial environmentalism since the post-war 'Atom for Peace' discourse; and colonial history of science and technology pertaining to ecology, economy and energy (Sato 2017; Bahng 2020). In November 2019, just before the pandemic began, I visited my grandparents in Japan. As I watched my children meet their great-grandparents, I recalled my own childhood memories. Growing up, I had spent summer holidays in the city of Iwakuni in

Yamaguchi prefecture. Surrounded by beautiful mountains and sea, I remember camping by the sea and waking up early in the morning to look for *kabutomushi* (rhino beetle) in the woods – rich sensory experiences my body memory still recalls. Iwakuni is also located just 40km from the hypocenter of the nuclear bomb dropped in Hiroshima. As a child, I listened to countless stories of war. I listened to accounts of nuclear weapon as real, lived experiences and why it must be never repeated. I remembered *obaachan* (grandmother) telling me her childhood memory during the war. She said she was taught to sharpen bamboo sticks for self-defence at school. Then at night, she would light a candle to read books – what she really wanted to do at school. Laughing at the absurdity of war, she said that seeing us being able to play by the sea and read books made her happy. Recalling my childhood memory and my *obaachan*'s childhood memory made me realize that the simple matter of everyday life – that the children could live and grow in close relation to nature and be empowered to think and act – had been a matter of care across the generations. As I sat beside my *obaachan* and my mother, with my third child on my lap in a line of four generations, I saw myself as part of this generational trajectory. I realized my research was located at a critical juncture between the surviving generations with living memories of nuclear bombs and the emerging generations facing a future beyond nuclear disaster.

The last translation I did as part of this research project was a video speech recorded by Rin (pseudonym) in December 2021. In this speech, Rin voiced her concerns about the vision of future based on fossil fuels and nuclear power. Explaining that the future she faces for the next 40 years is the decommissioning of the stricken nuclear power plant, she urged a reimagining of better futures. As Rin voiced her hope towards 'children of tomorrow', and hearing the generational echoes, I discovered who this research was really for – the children who are yet to come, from whom we borrow this earth.

Conclusion: What's next?

One thing I learnt about AR journey is that each arrival signals a new departure. Although my doctoral research project concluded in December 2021, at the time of writing, the collaborative inquiry with the co-researchers continues. In March 2022, we launched another collaborative video in timing with the eleventh anniversary of 3.11. Rin's video speech was in this film. Minami, the 2015 camp participant who went to study in Germany, also participated in this project. As of July 2022, we are working with the Fukushima students on an environmental awareness campaign that was initiated by the students themselves.

In doing this AR project, I realized that it is important to continue listening to the real-life stories of nuclear displacement and the emerging voices of critical hope. I especially found that applying a transnational grassroots perspective is crucial because the problem is experienced not only by the people of Fukushima but also the Australian Indigenous peoples, the Pacific Islander communities and other communities around the world. As a Japanese researcher living in Australia and located in the Asia Pacific region, I feel that I am well positioned to listen out for these voices. Moreover, I am interested to further investigate the interrelated themes of children, youth and ecological citizenship. In the face of complex socioecological problems that directly affect the futures of emerging generations, I believe it is necessary to listen to their concerns and hopes in envisioning more liveable futures.

More significantly, having experienced AR, and being a part of this dynamic and engaging AR community, I see AR as an increasingly relevant and life-giving research paradigm. As an early career researcher and an emerging action researcher, I commit to building on the rich legacies of AR with a view to one day also passing it on to future action researchers. I hope to continue to do research that acts and inquires with real people on a real problem to shape and imagine better futures for all.

Acknowledgements

I would like to thank the reviewer and the editor for their helpful feedback on an earlier version of this article. I would also like to express my gratitude to Prof Akihiro Ogawa for initiating this special issue and to Ai Ming Chow, Ed Hyatt, Asha Ross and Abbie Trott for the in-depth and honest discussions of AR PhD experiences that enriched the reflections presented in this article. This research was supported by the Australian Government Research Training Program Scholarship, the Asia Institute Graduate Research Award received in 2017, 2018 and 2019, and the Graduate Research in Arts Travel Scholarship received in 2017, 2018 and 2019.

References

3.11 Japan Nuclear Disaster Aid Association, 2016, '*Genpatsu ni tomonau hoyō jittai chousa*' [Report on state of recuperation activities after the nuclear accident], 3.11 Japan Nuclear Disaster Aid Association, Tokyo.

Bahng, A, 2020, 'The Pacific proving grounds and the proliferation of settler environmentalism', *The Journal of Transnational American Studies*, vol. 11, no. 2), pp. 45-73.

Bradbury, H, 2015, 'Introduction: How to situate and define Action Research' in Bradbury, H (ed), *The Sage Handbook of Action Research* (3rd edn), Sage Publications, Thousand Oaks, pp. 2-15.

Brown, TM & Dreby, J,(eds) 2013, *Family and work in everyday ethnography*, Temple University Press, Philadelphia.

Chandler, D, & Reid, J, 2016, *The neoliberal subject: resilience, adaptation and vulnerability*, Rowman & Littlefield International, London.

Clark, T, 2008, 'We're over-researched here!' Exploring accounts of research fatigue within qualitative research engagements, *Sociology*, vol. 42, no. 5, pp. 953-970.

Collier, S & Lakoff, A, 2005, 'On regimes of living', in Ong, A & Collier, S (eds), *Global assemblages: technologies, politics and ethics, as anthropologocal problems*, Blackwell, Malden, pp. 22-39.

Fisher, B & Tronto, J, 1990, 'Toward a feminist theory of caring', in Abel, EK & Nelson, MK (eds), *Circles of care: work and identity in women's lives*, State University of New York Press, Albany, pp. 36-54.

Frost, N & Holt, A, 2014, 'Mother, researcher, feminist, woman: reflections on 'maternal status' as a researcher identity', *Qualitative Research Journal*, vol. 14, no. 2, pp. 90-102.

Gilligan, C, 1982, *In a different voice: psychological theory and women's development*, Harvard University Press, Cambridge.

Greenwood, D J & Levin, M, 2007, *Introduction to action research: Social Research for Social Change* 2nd ed,, Sage Publications, Thousand Oaks.

Haraway, D, 2016, *Staying with the trouble: making kin in the Chthulucene*,Duke University Press, Durham.

Hikita, K, 2018, *Genpatsu jikogo no kodomo hoyo shien: "Hinan" to "Fukko" to tomoni [Recuperation support for children following the nuclear accident: Alongside both "Evacuation" and "Reconstruction"]*,Jimbun Shoin, Kyoto.

Latour, B, 2004, 'Why has critique run out of steam? From matters of fact to matters of concern', *Critical Inquiry* vol. 30, no. 2, pp. 225-248.

Middleton, J & Samanani, F, 2021, 'Accounting for care within human geography', *Transactions of the Institute of British Geographers*, vol. 46, no. 1, pp. 29-43.

Mikami, A, 2021, 'Translocal civil society and grassroots resilience; A case study of the Fukushima-Cairns recuperation initiative',. in Avenell, S & Ogawa, A (eds), *Transnational civil society in Asia: the potential of grassroots regionalization*,Routledge, Abingdon, pp. 23-39.

Moriggi, A, Soini, K, Franklin, A & Roep, D, 2020, 'A care-based approach to transformative change: ethically-informed practices, relational response-ability & emotional awareness', *Ethics, Policy & Environment*, vol. 23, no. 3, pp. 281-298.

Ogawa, A, 2009, *The failure of civil society? The third sector and the state in contemporary Japan*, State University of New York Press, Albany.

Ogawa, A, 2020, 'Envisioning anthropology of policy: Through civil society as an analytical lens – A keynote speech', in *International symposium of law, accounting and culture of nonprofits in East Asia – Universality and Individuality*, August 28, National Museum of Ethnology, Osaka, Japan.

Puig de la Bellacasa, M, 2017, *Matters of care: speculative ethics in more than human worlds*,University of Minnesota Press, Minneapolis.

Sato, K, 2017, 'Japan's nuclear imaginaries before and after Fukushima: Visions of science, technology, and society', in Ahn, J, Guarnieri, F & Furuta. K (eds), *Resilience: a new paradigm of nuclear safety: From accident mitigation to resilient society facing extreme situations*,Springer, Cham, pp. 195-206. https://doi.org/10.1007/978-3-319-58768-4_15

Tozzi, A, 2021, 'Reimagining climate-informed development: From 'matters of fact' to 'matters of care'', *The Geographical Journal*, vol. 187, no. 2, pp. 155-167.

Tronto, J, 1993, *Moral boundaries: a political argument for an ethic of care*, Routledge, New York.

Wadsworth, Y, 2014, *Building in research and evaluation: human inquiry for living systems*, Allen & Unwin, Sydney.

Wadsworth, Y, 2015, 'Shared inquiry capabilities and differing inquiry preferences: navigating 'full cycle' iterations of Action Research', in Bradbury, H (ed), *The Sage Handbook of Action Research* (3rd edn), Sage Publications Ltd, Thousand Oaks, pp. 750-759.

Biography

Akina Mikami is interested in how creative practices of research, art and activism contribute to socioecological flourishing. She completed her PhD at the Asia Institute, the University of Melbourne. Her doctoral thesis investigated how grassroots civil society shapes alternative resilient futures in disaster recovery politics. She is trained in interdisciplinary field of media and communication studies that build on anthropology, sociology, geography and philosophy of science and technology. Her research interrogates social-environmental change at the interface of agency, space and politics with a focus on young people and ecological citizenship. Akina also works as a Monitoring, Evaluation and Learning (MEL) specialist evaluating media development initiatives in the Asia Pacific region. Her work has been published as journal articles in international peer-reviewed journals including International Journal of Communication and Global Media and Communication and as a chapter in a Routledge edited book, Transnational Civil Society in Asia.

Conclusion: Walking beside co-researchers and finding off ramps in PhD dissertation journeys

Ai Ming Chow, Ed Hyatt, Akina Mikami, Asha Ross, Abbie Trott

Abstract

As the contributing authors of this Special Issue: PhD Journey in Action Research, we conclude this issue by offering a final discussion on doing Action Research (AR) in the context of doctoral research. Tracing our collaborative reflection and writing process, we discuss a range of questions that we felt might be helpful for new students to embark upon their AR journeys. These include: how AR is different from more conventional research; how research ethics can be handled; how AR projects can fit into a PhD timeline; and how the philosophical principles of AR are understood and expressed in doctoral dissertations. Situating our collaborative discussion within the dynamic and expanding field of AR, we suggest that our individual AR PhD projects and our collective reflections directly partake in the making of a new AR stream.[1]

Key words: Action research (AR), AR PhD, AR thesis, collaborative reflection, collaborative writing, triple loop learning, recoverability, co-generation

---

1 Authors are listed in alphabetical order. All authors contributed equally to this work.

What is known about the topic?

There are existing discussions on the topic of Action Research (AR) conducted in the context of doctoral research (Dick, 1997; Gibbon, 2002; Greenwood & Levin, 2007: 251; Herr & Anderson, 2014; Klocker, 2012; Macguire, 1993; Ogawa, 2009; Southby, 2017). The topic of collaborative reflections, learnings and writings in AR has also been discussed (Austin, Bessemer, Goff, Hill, Orr & Varrtjes, 2021). However, a collaborative reflection by AR graduate researchers on their PhD experiences and what lessons can be useful for future AR students has not been completed.

What does this paper add?

This paper offers a collaborative reflection of AR PhD experiences by discussing a range of questions including: how AR is different from more conventional research; how research ethics can be handled; how AR projects can fit into a PhD timeline; and how the philosophical principles of AR are understood and expressed in doctoral dissertations. The paper also offers strategies the authors found useful that current and aspiring PhD students can apply to their AR research.

Who will benefit from its content?

- PhD students potentially or currently doing AR as part of their research project
- Academic staff in who are potentially or currently supervising an AR PhD project

What is the relevance to AL and AR scholars and practitioners?

- This paper discusses the collaborative process of reflecting, learning and writing about AR PhD experiences.
- This paper considers various aspects of applying AR principles to doctoral research projects, such as differences in conducting more conventional research, research ethics, timeline and the diverse forms of AR dissertations.
- This paper demonstrates that AR principles and values can be translated into and expressed in diverse ways across different doctoral research projects in different disciplines.
- This paper raises further questions about the opportunities and challenges of doing AR PhD at various levels including the nature of research projects, relationship with research participants, the receptiveness of AR by supervisors and departments, institutional structures around ethics and candidature, and the broader changes in higher education.
- This paper situates specific AR PhD experiences in the broader developments of an AR stream unfolding in Melbourne, Australia.

Received   June 2022   Reviewed   July 2022   Published   November 2022

Introduction

As we discussed and reflected upon our doctoral research projects throughout 2020 and 2021, we found that we were often confronted by a series of questions. These questions drew our research together across different projects, approaches and relationships to action research (AR). These questions include: how AR is distinct from more conventional research, how ethics in a university setting can be handled, how AR projects can fit into a PhD timeline, and finally, how the philosophical underpinnings of AR are understood and can be expressed in a PhD thesis. This conclusion is structured to offer our collective insights on these questions.

To answer these questions, our approach was to use a co-writing model that emerged as we drafted and revised our articles. Similar to Austin, Bessemer, Goff, Hill, Orr and Varrtjes (2021), we engaged in a writing process as collaborative authors who move between writing, reflecting, reviewing and shaping our thinking through multiple cycles. The process of writing our articles spans a number of stages. All five of us participated in the unit *Introduction to Action Research for Graduate Researchers* coordinated by Akihiro Ogawa. Ming, Ed and Akina were in the first year the subject ran, 2016, and Abbie and Asha were in the second class in 2017. Later, each of us presented a seminar about our AR experiences to subsequent cohorts of that same subject. Ming, Akina, and Asha presented to the 2018 cohort, Ed and Abbie to the 2019 group, Akina and Asha to the 2020 cohort, and then Abbie and Asha to the 2021 class.

The next stage of our journey was to present at a two-day workshop late in 2020. At this mini conference, we each delivered a 20-minute paper, from which we received feedback from the group. Asha, Abbie, Akina and Ming presented here, and Ed attended as an active listener. Abbie based her paper on the seminar presentation she had given in 2019 but expanded her ideas. Akina focused on field experiences of collaborative inquiry, which she had found garnered the most interest from the students she presented to in 2018. Ming offered her reflexive account of conducting research on Indigenous art and cultures as a non-Indigenous researcher. She also presented her emerging ideas on the parallels between an AR paradigm and the decolonisation movements. Building on these presentations, in April 2021 we each submitted an 8,000-word essay. We then met over two days to discuss each essay and provide feedback. After this, in late 2021 we each gave feedback on the next draft of one other person's essay. Not everyone received feedback, or was able to provide a draft, but the notion was that we would engage in some collaborative editing before the final drafts made it to Akihiro in January 2022. It is important to note that our colleague Asha Ross was not able to submit her final paper after the first stage of review, but her work was integral to the co-writing of the project. For that reason, we have left her work here in the conclusion where she offers her reflection and insights.

August 2021 is when we began co-writing this Conclusion. First, we established the five questions we thought would hold our articles together. Next, the five of us spent time individually thinking about our responses and recorded an online Zoom conversation where we discussed our responses to each prompt. From there, each person took responsibility for transcribing the responses to one question and writing them up into a 1,000-word response. We then undertook a co-editing process where each person revised the entire conclusion until the final draft was ready for journal submission.

The process we used was not drawn explicitly from other ways of co-writing that have been discussed in literature. Some of us thought of it as an edited interview such as the one done in conversational style by Aftab Erfan and Bill Torbet in their discussions on Collaborative Developmental Action Inquiry (Erfan & Torbet 2015). Henry Jenkins, Mizuko Ito and Danah Boyd's edited interview across *Participatory Culture in a Networked Era: A Conversation on Youth, Learning, Commerce and Politics* (Jenkins et al., 2016) is another example. Upon reflection, we realized that it would be more useful for our intended audience of emerging AR scholars to turn our conversations into prose.

In their article, *Collaborative writing as action research: A story in the making*, Diana Austin and colleagues (2021) describe their AR process of collaborative writing as being integral to their research into how AR manifested more broadly through the publication of AR knowledge. The authors captured their process in real time, the results of which they communicate through a narrative exploration of the process. This framing of narration in AR is an apt description of how we have described our AR process here and identifies the importance of reflexivity for knowledge generation and the haphazard nature of our co-writing process.

The process we used could be described as Collaboration 3.0, which Schimmer argues is generative and innovative, where the problem needs a solution that 'is too challenging to solve alone' (Schimmer cited in Austin et al 2021, p. 41). As a group, we needed to work together to generate the knowledge required to make sense of the questions we were asking of ourselves on behalf of our imagined audience. Austin and colleagues reflected on triple loop learning as an alternative way to understand Collaboration 3.0 in response to a reviewer suggestion but felt that it took away from their sequential thinking. In a critical review of triple loop learning, Paul Tosey, Max Visser and Mark NK Saunders (2012) propose that while triple loop learning has been attributed to Chris Argyris and David Schön, more rightly, its origins are in the work of Gregory Bateson.

Focused on the different processes of organizational learning, Bateson (1987) described the third order of learning, or Learning III, as "a corrective change in the system of *sets* of alternatives from which choice is made" (Bateson, 1987, p. 298, emphasis in original). Tosey, Visser and Saunders' (2012) research indicates that Bateson's notion of Learning III offers an important theoretical foundation to the idea of triple loop learning. First, they note that Bateson's scepticism about the instrumental pursuit of Learning III is important because it questions whether generative learning can or should be fully predictable and controllable. Second, they suggest that Bateson's idea of Learning III extends beyond language into aesthetics and the unconscious is noteworthy because merely talking about the learning processes does not constitute Learning III. Third, they observe that Bateson's description of *recursive* nature of learning, rather than a hierarchical one, is crucial because "higher orders of learning are not inherently superior to or more desirable than lower levels" (Tosey et al 2012, p. 299). Finally, they discuss the importance of Bateson's caution around the risks of Level III, highlighting the ideological dimension to learning that questions in whose interests the organizational learning is being driven. For Robert Flood and Norma Romm, triple loop learning steps beyond the imperialism, complementarianism, pragmatism and isolationism of Critical Systems Thinking (CST) into a postmodernist frame where inherent research tensions are articulated through active choices, which they frame as diversity management (Flood & Romm 1995). Therefore, triple loop learning offers strategies to manage a diversity of issues and dilemmas. Attenuating single and double loop learning, triple loop learning draws three central questions together: 'Are we doing things right, are we doing the right things, and is rightness buttressed by mightiness or mightiness buttressed by rightness?' (Flood and Romm 2018, p. 266) – how, what and why in all loops, reflecting Bateson's model of third order learning (Tosey et al 2012).

Reflecting on this ongoing process of collaborative writing, what triple loop learning offers us are strategies to navigate the dilemmas and issues that are present because our practices as emerging AR researchers are diverse, and often contradictory. In retrospect, we were able to concurrently address the overlapping how, what and why of our projects' intersections and find the overlapping concerns that hounded our research. Moreover, as the co-writing progressed, we were able to also navigate the diversity of our voices reflected in the form of varied tones in a co-written work. Upon discussing whether to streamline our tones or keep them distinct, we decided to shape our collaborative voice through iterative process of collaborative editing. As it went through many rounds of editing, we interestingly observed that the variety of tones were sustained but the jarring aspects of it faded away. In doing so, our collaborative voice became more coherent without losing the diversity. As such, we were able to tell our story/ies where the multiple reflections were co-creative, and the relationality across the networks of our work and ideas shone through our collaborative voice. Here, that is evidenced through the active naming of ourselves as participant/researchers in the PhD AR journey. The generative nature of how we developed our questions and answers is indicative of what was 'known at that time and in this space' (Austin et al., 2021, p. 45).

In this co-generative process of reflection and writing, the five questions that we saw as holding together our AR journey, and of being of some use to future AR postgraduate students, were as follows:

- How is AR practically different to other more conventional research?
- How did you handle ethics in your AR research?
- How can an AR project be done within a PhD timeline?
- How did you think about the philosophical underpinnings of AR as a paradigm?
- Reflecting back, how was your philosophical approach to AR brought into being in your project?

We present the co-generated reflections on these questions below. We hope that what we learnt is useful to other doctoral candidates as they make their way through the AR journey.

How is AR practically different to other more conventional research?

In our conversations, we discussed how AR is practically different to more conventional research, particularly in terms of conducting fieldwork. A caveat worth mentioning at the outset, is that admittedly none of us have conducted much conventional research and our dissertations represent our largest individual research projects to date. Therefore, we only have some experience of fieldwork, and our ability to compare these types of research approaches to ethnography is limited. However, we did seem to rapidly settle on what we perceive to be three core differences between AR and more conventional research: the level of objectivity or subjectivity in the researcher's position, the boundaries with research participants, and the desired outcome of the research.

Objectivity vs subjectivity

The question of objectivity in social science research is not a concern reserved solely for AR. For example, Julia Scott-Jones refers to her earliest exposure to ethnography and social science research, where the researcher might 'over-identify with the research participants and lose all sense of objectivity' (Scott-Jones, 2010, p. 3). But, as with most research disciplines, social science research has changed and objectivity has given way to subjectivity (Scott-Jones, 2010, p. 5). As Ming commented, an AR researcher immerses themselves in the research process rather than remaining outside it. Asha noted the relief she felt in the field, knowing that she did not have to be objective in her research role and that she could embrace being subjective in order to engage directly with her research challenges. Akina remarked that there is a very specific, real-life context to AR that is not abstract or entirely theoretical like it can be in conventional research. This means the researcher cannot be entirely objective. The researcher's own views will almost certainly change, even if only subtly, from engaging with a community. Ed likewise felt that AR is not intended for researchers to take an objective stance, that the researcher purposely embeds themself, and therefore remains open to altering their view as the research evolves. So, the position of the researcher is necessarily subjective rather than objective from the outset and continues to be so as the research progresses. This tracing of the subjective viewpoint is a hallmark of AR.

Not only is the research process going to alter the research participants, but it will likely alter the researcher too. This is because engaging directly with real-world challenges requires a commitment to the experience. The researcher should therefore be open to having life-changing experiences in the field and be prepared for their positionality to naturally evolve based on what they are exposed to in the process. AR involves deep, embodied experiences which will change the researcher because they are so involved in what is going on, and presumably care about it in some way. Therefore, it would be surprising if there was no impact from being involved with helping people effectively change some aspect of their life.

Boundaries

A researcher's sense of their own subject position and level of objectivity in the process also means that the boundaries with research participants are different in AR. AR aims to generate knowledge-making through relationships with research participants, thereby treating them as co-researchers. Akina expressed that in her research, 'research participants are co-researchers, not just research subjects or research objects'. In conventional research a separation between researcher and research participants is desired, but this is the opposite in AR. In AR, the researcher typically develops intimate bonds with the community. As Asha noted, you are just one person amongst many, and necessarily not the most important person. AR pulls together a number of voices in the experience of change, and boundaries are drawn in different places as that experience progresses.

Abbie also felt that due to the more inclusive nature of AR, making participants co-researchers in the process, there is a lot of additional work on the researcher's part to engage with participants during the research process in such a way that they have ownership of the research. Asha also noted the higher potential for conflict or unforeseen problems in fieldwork because the researcher becomes a new element in the context of the challenge facing the community:

> "You are committing to being involved. And it is not just your involvement in how you read a situation. You are encouraging other people to identify how they read the

situation. So, you are not just observing what is going on, you are also trying to get people to be engaged, to be involved, to be active. I feel like there is a lot more operating in grey areas."

Change as outcome

The desired outcome of AR usually involves some meaningful change in the community or in the lives of people who participated in the research. Ming noted that an AR researcher is often trying to do more than just capture what is happening; there is an attempt to foster change because of the research. This is distinct from the approach utilized by researchers in conventional research who seek to maintain a sense of objectivity and therefore maintain more distance from the research participants, but also to minimize their own influence on the community and its future.

Akina commented that in her field there are two distinct streams of conventional research. One stream contains a strong applied focus, and some of this is done in the vein of AR, but knowledge tends to be rendered technical and the nuances of everyday engagement or tensions are often lost. The other type of conventional research is more theory-oriented, where "the problem is raised from theory and the purpose of it is to advance more theory. So, it is just theory without engagement with real-life contexts". She felt her dissertation was an attempt to draw on both of those existing streams of research but to still try to achieve a true action-research attempt to foster social change.

The reflections articulated above are not exclusive to AR but are issues considered to some extent by researchers in all types of research. Through our conversations about AR, we concurred that these considerations were central to defining what AR is and how it can differ from the nature of more conventional research. Identifying those differences is key to aiding in the progression and integration of AR into the ecosystem of research. These differences also suggest that there are implications for ethics and logistics, the topics of two other sections in this Conclusion, that go beyond the concerns of conventional research. For example, AR is often open-ended, and the boundaries are less fixed, which often leads to an ongoing project and involvement in the community. This results in less of a sense of completion which can be more easily achieved with a 'one-and-done' survey. Another difficulty arises from being embedded in the community as an AR researcher. It can be challenging to balance how you conduct yourself, and how much you share your opinions during the study because you might be a future, ongoing member of that community after the formal investigation ends. So, there is a balance in honouring your immediate research needs and who you present yourself as in the community with any future relationships you anticipate having within that community. These types of issues are discussed in greater detail in other sections of this Conclusion because they are especially important for future PhD AR researchers to consider.

How did you handle ethics in your AR project?

During our discussions, we often circle back to the issues related to ethics and the recurring question remains: How do we handle ethics in conducting action-oriented research projects? Our discussions revolved around the process of getting ethics approved by an ethics committee at the university, how do we maintain ethical conduct as action researchers, and how do we reflect carefully on our own ethical stances towards engaging with various groups of stakeholders. In Davison, Martinsons and Wong's (2020) review of AR literature, they found that ethical dilemmas and their resolutions within action-oriented research were addressed superficially, without sufficient details provided. While it has been acknowledged that 'action research combines theory and practice (and researchers and practitioners) through change and reflection in an immediate problematic situation within a mutually acceptable ethical framework' (Avison et al., 1999, p. 94), there is limited discussion on such an ethical framework. In addition, Greenwood (2002) also identified ethical concerns as one of the main challenges that action researchers need to confront and wrestle with. Therefore, in this section, we aim to highlight the few key themes that emerged from our individual AR projects as we reflect on how we dealt with ethical issues.

Action researchers strive to work collaboratively with an organisation or community as partners which create the expectations that both parties are actively and interactively involved (Brydon-Miller, Greenwood & Eikeland, 2006). Davison et al. (2020) suggested that there is a need to formalise a set of ethical principles and criteria, in which ethical issues are reported consistently and in great detail. These should form the ethical framework of the project which is expected to be accepted and agreed upon by the researchers and the partner organisations and communities involved. As doctoral students conducting AR research projects, we too had to provide detailed ethical frameworks by outlining in detail the actions or plans put in place to ensure that the participants, the researcher and the data collected are treated with the highest ethical conduct.

One of the key considerations that is consistent across our projects, is the extra care and steps needed when our projects require participants who are considered part of vulnerable groups. These include children and/or young people (<18 years old), Aboriginal and/or Torres Strait Islander individuals or peoples, or people with cognitive impairment, intellectual disability or mental illness. For example, Ming's research involves engaging closely with Indigenous business owners and artists, Asha and Abbie interact with high school children and young people below 18, and Akina deals with children who have experienced nuclear accidents. We acknowledge that extra thought and care is needed when ethical issues arise to ensure that the rights of the participants are protected.

As a result, we devised ethical frameworks and strategies for our projects when conducting research with these participants that are considered vulnerable. For Ming, she ensured that she contextualised herself with the history of Indigenous people, arts and cultures in Australia and to *decolonise her minds* (Ngũgĩ wa Thiong'o 1986,; Tuhiwai Smith, 2012). Before conceptualising her research project, she immersed herself by spending time with the local Indigenous communities living in Ntaria, having conversations and meetings with local leaders and Elders, to understand the impact of research for the Indigenous communities as well as the outcomes which the communities seek to achieve. These interactions and conversations with the local Indigenous members ultimately guided and shaped her research aims and goals. Additionally, when putting together her ethics application, she explicitly declared that she is collaborating with Indigenous art business owners, regardless of whether they are Indigenous or not. This stance helped her to challenge her colonial biases that tend to perceive Indigenous people as 'vulnerable'. Instead, based on her interactions with the local Indigenous communities, these groups demonstrate great resilience and insights. Especially through the AR research approach, the Indigenous participants are the ones who are guiding the process and making the calls, working as collaborative partners. This is also because action researchers do not exercise control over all the activities of a project (Davison et al., 2004).

As for Asha and Abbie, they both agreed that when conducting research with the high school children and young people, they needed to seek consent from either the principals of the schools or the parents before conducting any interviews. More importantly, Abbie and Asha ensured that, when speaking with these younger participants, they would downplay their expertise in the project. Instead, despite being 'vulnerable', Abbie and Asha extend great respect for the participants' perspectives, ensuring that the young participants are positioned as the experts, and that they could walk beside them. Abbie reflected that this was an ethical choice to empower the participants towards being action-oriented. When engaging with the younger participants, Asha also explicitly explained,

> "it wasn't really about me or the universities, but about what do the young participants think research can learn from them [young participants] ... I wanted to understand what parkour vision was for them [the young participants] so it wasn't about me coming in to tell them how to solve their problems. But it is more letting them know that they have something that we [researchers] should be paying attention to".

Such a stance is aligned with Davison et al. (2020) who argued that both researchers and participants need to consider the ethics of their roles and responsibilities, so that they are motivated to enact behaviour that is in their own best interests, which would lead to outcomes that satisfy both researchers and participants.

Lastly, Akina provided further insights from her research with a civil society organization that supports children who experienced nuclear disaster. Although her co-researchers were volunteers of this organization and therefore not considered vulnerable, she interacted directly with children aged between 12 and 15 through her involvement as a volunteer. Therefore, when developing her ethical framework, she took seriously the historical and political context of disaster recovery and prioritized the agency of her co-researchers (volunteers) and their participants (children). In her ethics application, she clarified that her main research partner was the civil society organisation. However, also given her role as a volunteer, she provided detailed plans to ensure that the children's rights and welfare were protected. This entailed providing strategies such as a distress protocol (Haigh & Witham, 2015) and becoming well-versed with any triggering words. In dealing with the 'ethics of collaboration' (Davison et al., 2020, p. 7-8), Akina engaged with her co-researchers as peer volunteers to centre the voices of children in the research process from identifying a question, planning, acting and evaluating. In this process, the children were addressed not as victims or merely survivors but as unique individuals, storytellers and change-makers. Akina also regularly presented her thesis, as it developed, to her co-researchers and invited them to her completion seminar to incorporate their feedback before finalizing her thesis.

Overall, our experiences in dealing with ethics demonstrate the importance of providing detailed plans and frameworks about our positionalities and our assumptions as researchers. In the process of conceptualising these ethics frameworks for the project, it has helped us to reflect on how our assumptions drive our actions and goals, which would then dictate the changes we see over time. Additionally, we acknowledge the principles of research ethics that include respect, beneficence and justice that would ultimately guide our research approach. This is because for AR projects, the process is often dynamic and interwoven throughout the ongoing action-reflection cycle, which calls for AR researchers to be continuously vigilant and open to new ethical considerations through ongoing reflexivity.

How can an AR project be done within a PhD timeline?

AR is time-intensive, even by the standards of conventional research. There is, or should be, a reasonable concern among PhD students for whether and how an AR project can be completed within the time parameters of a dissertation program. As a group, we felt that you can certainly start an AR project in a PhD program, but you likely cannot finish the entire project during this time. This was not a gloomy appraisal; we felt that this was fine as long as a student's mindset was calibrated for this reality. We all agreed that it is naturally difficult to incorporate a loop-research mindset into a linear process that is expecting a specific product (the thesis). As Ming put it, it is "a balance between what you plan to do and what you can practically do, especially because it can be a struggle to do the AR loop, since this is a linear process. There can be a tension there." To help adopt the necessary mindset to manage this natural tension, we concluded that there are two key things that an AR PhD student should aim to do: build modularity into the AR project and operate as a project manager.

Modular

All of us felt that an AR project is more of an ongoing effort, not just an isolated project like most conventional research. Because this is the case, it would behove a PhD student to design the overall project in modular fashion to emulate conventional research. You might need distinct studies (or distinct parts of the thesis) that can be completed to show your examiners that you can conduct research, the ultimate point of a thesis. Several of us liked the concept of thinking of the studies as off-ramps on a highway (so much so we included it as part of the title for this conclusion). One of these off-ramps might be well suited to satisfy the requirements of a PhD program, leaving the highway to continue as the backbone of a larger research agenda. In this way, an AR project is never 100% done, but it can be made provisionally complete enough to earn your doctorate. To develop off-ramps, ask yourself: If you suddenly only had six months left to complete your thesis, what would you cut out or how would you reframe your study out of necessity? What if you had one year remaining, then what would you cut out? One of these off-ramps may turn out to be the one that allows you to turn in a finished product on time.

Similar to more conventional dissertations, it is almost certain that you will have to put aside some material that seems very interesting to you, just for the sake of focusing on the material required for thesis submission. For intellectually curious people who enjoy digging into the thickets of life in rigorous fashion, this can be difficult. Several of us commented that we keep a running file of ideas that can be referenced later if/when we have the space to do so. As Asha mentioned, you should "start directing some of this stuff into other areas, and then clarify and clean it up and you can revisit areas that are really interesting later." Over time you may lose interest in the idea, or discover that it has been done already, or you will have an ideal head-start on a new branch of your research agenda.

In regard to a PhD only ever being provisionally complete in the larger sense of your research agenda, you should know exactly how provisionally complete it needs to be in order to get it done. What are the hard-and-fast requirements that need to be satisfied in order to complete the PhD process? Then you can look for ways to facilitate meeting these requirements, or to reach that target off-ramp with a thesis deemed complete by the powers-that-be. Perhaps align your project or some piece of it with something that was already happening in the community or moving along anyway. Asha felt that she was able to make a film within her allotted PhD deadlines only because her parkour group had already intended to make such a film and she came along at the perfect time. Abbie did the majority of her reporting back to the community only after her writing had been completed, similar to Ed and Ming. We all feel strongly that you can fulfill the spirit of an AR project and finish your thesis even if you cannot complete all the desired or originally planned activities during the PhD timeline.

Project manager

We also felt that it is advisable for an AR PhD student to approach their thesis as a project manager, employing practices that will increase your odds of successfully completing it with minimal angst. To that end, as a group we discussed several practices that every PhD student should consider engaging in regardless of whether they are completing an AR or other research project:

- Make an agenda for every supervisor meeting;
- Keep notes about what was agreed upon in those meetings and send these to the other participants shortly afterwards so that there is a record;
- Track everything in time – by when is this to be done, and by whom?;
- Work with the next milestone in mind, designing milestones by working backwards in time from the final goal;
- Build in extra space for inevitable breakdowns or delays;
- If possible, get confirmation from different sources on critical items (i.e., triangulation);
- Know the requirements of your program yourself – do not assume your supervisors know them.

As described by several of us during our conversation, you have to accept the situation that you are still a student, you are doing a PhD, you are operating in a system, and you need to jump through hoops to move on. Abbie neatly summed up our collective thoughts on the link between project management and a PhD journey:

> "You have to use your project-management skills. Your heart and soul might be in the research, but it is a project. And the purpose of a PhD is to write 80,000 words about a thing. That is the task, you just have to do the task. That is the project. No matter how [much] you love it or are invested in it, or how much you hate it and are no longer invested in it, it is a piece of work you have to do to get a 'Dr'. next to your name."

Akina can offer a unique perspective on the project management aspect because she is doing a PhD part-time while raising her family, making her project duration the longest out of all of us. As she pointed out, the university PhD timeline assumes a free, independent individual with no caring commitments. This can conflict with anyone wanting to raise a child, care for an elderly person, or provide support to someone who needs additional assistance in life. She acknowledged some benefits from her extended timeline, such as how she has been able to develop stronger relationships with her research subjects. But similar to the rest of us, she has also had to draw the line at what will be included in her thesis because otherwise, as she stated, 'it could be endless'.

A concept that could be usefully applied to the potential endlessness of an AR project is the idea of 'recoverability' (Checkland and Holwell, 1998). Recoverability refers to the ability of outsiders to more rigorously scrutinize the judgments and research process used by a researcher based on an epistemology that the researcher has declared in advance. Generally, this concept offers AR researchers a way to legitimise AR by approximating the concept of replicability that is often paramount for perceptions of rigour (Checkland & Holwell 1998). What recoverability also offers PhD candidates is a way to know when to declare the AR project provisionally complete for thesis purposes because what will be considered acquired knowledge has been declared in advance. Thus, the project not only has additional rigour, but it will not be an endless venture.

It appears that no matter the length of time in your PhD program, all students will experience a similar challenge of having to draw boundaries around the scope of their PhD. But take heart; this is a manageable challenge, hopefully made more manageable with thesis modularity and a project management mindset. As Akina eloquently put it, you should consider the PhD as "one song in a full album that you make throughout your career".

How did you think about the philosophical underpinnings of AR as a paradigm?

When thinking about how AR had affected and effected our research projects, we realised that central to how we thought about AR in relation to them, and a commonality across the work that we were doing was that it was a mindset. The philosophical nature and essence of AR was at the heart of each of our projects, even though they were not all AR in the traditional sense. In describing how she sees AR paradigmatically, Akina wrote that she finds that AR is a relational, situated, contested, dynamic, reflexive, generative and transformational journey. We find these are useful ways to arrange our musings and thoughts as we try to answer our fourth question: How did you think about the philosophical underpinnings of AR as a paradigm?

For Akina, *relationality* is the ontological and epistemological premise of AR. This is the relationality between researcher, and participants, and a relational approach to knowledge generation. By thinking about the relationality of the research, we were all able to understand the research, the participants, and the knowledge they generated in different ways. For example, Ed said of his research that:

> "I certainly approached it as if the participants themselves were also informing me about what was going on rather than the other way around. And that definitely shaped some conversations. I know I got information out of people that they wouldn't have ever otherwise shared in interviews and such because of that."

By *situatedness*, we mean contextual specificity. Constantly reminding ourselves that we are dealing with real people in a real-life context. As researchers we came to understand that we all existed in communities, the community that we were researching, the community that we came from, and the community that we were going to. Asha described this when she said of her research that:

> "my mindset was really connected towards my positioning within a community and so I wanted to position myself as a facilitator, or not leading, not driving, but rather, sort of in the middle of a bunch of things and looking around trying to see what people were saying and as accurately as I could portray their perspective and how it was feeding into the larger community perspective."

Paradigmatically, AR is also *contested*. Because AR is about researchers and participants inquiring together and generating knowledge together, and because AR brings theory and practice together, that knowledge is always negotiated and contested. A range of related concerns that are entangled with the problem are revealed through an AR process. As a result, what we want, or hope, or even think might emerge from the research is conflicted, as opposing knowledge shuffles for space in the generative process. For Ming, it involves 'unlearning' what she knows about how to solve the problems for the Indigenous art communities. Very early in her research, she learnt that the Indigenous art producers and artists are the custodians of their stories, and they are the one who make decisions about what is best for the communities, not the researchers involved.

The *dynamism* of research, and AR in particular, stood out for us when we were talking about the philosophical underpinnings of AR. This might be as living systems, loops, spirals and assemblage. For Akina, it was a way of engaging with the world of research as living, real and active with distinct rhythms and temperatures. Applying assemblage thinking and the idea of living systems to her project, she found that approaching research as inherently dynamic gave her and her co-researchers a breathing space to collaboratively make sense of the contingent, heterogenous and emergent components that were always on-the-move and in-the-making.

The *reflexivity* of AR goes more than one way. We must be reflexive about how we are generating knowledge; we must also be reflective in how we share it. Ed described this as speaking "the language of the paradigm that I'm in." In order to communicate his findings, he needed to strip the language of AR out, and use the language that his readers would understand to describe AR concepts. Within the context of generating knowledge, Asha found that she needed to "embody the nature of the spiral … to reflect and to reposition as required, based on the actual circumstances you are involved in, as opposed to what you expected the circumstances to be, what the university expected the circumstances to be, what you needed them to be for the project or anything like that, it's actually engaging in this living system."

Generating knowledge, the sixth node we are describing here, is ongoing in opening up to something new; leading to new questions and new possibilities on which to act and reflect. It is always followed by 'What's next?' For Akina, this manifested in her research by the sense that she was walking beside her co-participants. "It's staying with the trouble", she said, "acting alongside, you're almost creating a 'third space' where different conflicts and tensions can co-exist at the same time but you're resisting the temptation to make sense of it all or making conflict out of it, but just sort of letting it unfold and staying with it."

Finally, the paradigm of AR we all worked towards is that it is *transformative*. For Abbie this means that AR changed her as a researcher because it "influences how you change what you are doing or how you grow as a researcher or how you exist within a living system." AR gives the researcher a framework to change in ways that conventional research does not. This includes the transformation of not only the research project but also the researchers themselves.

The philosophical underpinnings of AR have become a neat way to frame and inform the approach Abbie now takes as a researcher. She is interested in co-design, working with her participants, and the systems and lifestyles of research, of life next to research, and most importantly, of knowledge generation. As Ed said, "AR definitely influenced me, and it influenced my mindset, and how she approached things but it had to be much more in the how I did things." As we are still developing our understanding of AR, we are reflecting on it, and us, and our understanding of it is continuously changing.

Reflecting back, how was your philosophical approach to AR brought into being in your project?

Tracing the ways that AR came alive in our projects, our works individually and collectively suggest that AR PhD journeys are characterized by multiplicity, heterogeneity and potentiality, each with its unique challenges, opportunities, learnings and new questions. The constellation of our diverse AR experiences evokes an AR assemblage, a contingent ensemble of heterogeneous elements that manifest in different expressions of PhD journeys. As Abbie pointedly noted, "our projects can have different amounts or parts or elements of AR, resulting in diverse forms."

Reflecting on how the philosophical approach to AR is reflected in initiating the inquiry, Abbie commented that "to be a *true* AR it needs to come from the community" (emphasis added). In her view, for a project to be considered fully AR, the community has to approach the researcher with a question or a problem and suggest doing research together. "If you are not in that situation", says Abbie, "you have to just apply the (AR) paradigm". In response, Asha considered a perspective that "there is always somewhere you can help people without it being an outsider-saviour thing or without it being like you've got solutions". To Asha, initiating an AR project feels more like, "You guys (co-researchers) are doing something really interesting. And I would love to be a part of it, if I could". This reflection inspires further conversations on what counts as 'true' AR – whether there is one, and how such an ontological question shapes our views on what it means to initiate an AR project.

It is fair to say, however, that all AR projects begin with questions that call for both inquiry and action, and such was the case with all our projects. For Asha, it was witnessing the intriguing phenomenon of parkour. Drawn to the fascinating ways that bodies were moving through public spaces, she began doing parkour herself. Noticing that the parkour community was asking how they see themselves and how they want to be seen by others prompted Asha to embark upon a collaborative inquiry with the parkour community in Edinburgh, Melbourne and Tokyo around the theme of Parkour Vision. In Ming's case, she was attracted to the dynamic Indigenous art markets in Australia. Realizing that non-Indigenous stakeholders continue to have a strong presence and recognizing the colonial-settler history, Ming decided to do research that reimagines marketplaces so that Indigenous people and their cultures continue to flourish. She especially became interested in the commercial art intermediaries that hold influential positions in the marketplace and how they interact with the Indigenous artists.

The idea that research practice itself contributes to agency and empowerment was also important to Abbie, who investigated how young people engage with theatre in the digital media environment. Acknowledging the barriers young people face as theatre audiences, she wanted to do research that centres the young people as active participants who generate meaningful knowledge in theatre experiences. As for Ed, the question emerged while reviewing literature with an action mindset. He encountered a longstanding knowledge divide between academic research and real-world practice over the use and validity of unstructured interviews in employment processes and noticed the lack of practitioners' perspective on this problem. Aspiring to do research that solves real problems, Ed began asking whether more useful insights might be gained if the same problem was examined from the practitioners' viewpoint. For Akina, who was exploring how civil society enables resilience-building in disaster recovery, she felt stirred by 'recuperation practice,' a grassroots endeavour that supports children affected by nuclear disaster. Getting involved as a volunteer, she realized her co-researchers were asking how to rebuild better futures that centre the concerns and hopes voiced by Fukushima children. This then became a collaborative inquiry of how resilient futures can be reimagined at the grassroots level.

As our research journeys unfolded, the values, principles and components of AR took various expressions in our projects. In her project, Asha feels that the AR element of knowledge co-creation featured prominently. She found that her role as action researcher was to become a facilitator of knowledge generation, a process she believes was enabled by the strong relationships she developed with her co-researchers. Looking back, Asha saw the visual cue of a spiral (cyclical process of planning, acting, observing and reflecting) characterizing her project, which guided not only her research strategies but also her thesis format. Throughout the research, Asha was aware that her research was designed, conducted, and written as an entirely AR project.

Similarly, Abbie thinks the co-generative mode of knowledge formation was applied to her project. She found the idea of knowledge-cogeneration as living systems was salient in her analysis of how young people engaged with theatre performances. Upon reflection, Abbie wonders whether the knowledge co-creation element was something she already embodied as an educator and a theatre professional in her engagement with her community and students prior to encountering AR. Whether the AR elements were something already within us that were called forth in specific ways by doing AR, or whether they were something new that we adopted, or whether it was both – is an interesting question for further deliberation.

In Ming's case, she feels that it was the AR element of knowledge democratization that manifested in her project. Recognizing the uneven power relations in knowledge construction and the interrelated colonial history, she approached research as knowledge decolonization. Ming found that crucial to this whole process was reflexivity, a key element of AR, which uncovered her distinct positionality to examine the Australian indigenous art market. She especially reflected on the porous nature of the outsider/insider boundary as a non-Indigenous outsider, a Western institution-trained researcher, and a third-generation Malaysian Chinese person growing up in postcolonial society with a shared history of British colonial administration.

For Akina, she thought the AR elements of tackling real-life problems and co-creating knowledge foregrounded her project. She found the ideas of collaborative inquiry as living systems and co-generation of action narrative were prominent in her research. She felt that her role was to become a co-subject in action-reflection cycles with her co-researchers in ways that lead to next actions, while also advancing theoretical debates and challenging policy discourse. She admitted that due to her limited knowledge of AR, it was only in the later stage that she came to see her project as AR. She believes more examples of different and diverse AR PhD projects can help encourage students to shape their own AR projects.

Meanwhile, for Ed, the AR values of research as solving real-life problems and improving human condition were expressed in his project. Although he described his thesis as not fully AR because it does not entail emancipatory quality, iterative action-reflection cycles or interventions with community involvement, he is confident he adopted an AR mindset. Despite confronting indifference towards AR in a heavily psychometric-dominated field of human resource research, Ed exercised his problem-solving skills by redesigning his project to satisfy discipline-relevant requirements while also embodying the AR spirit. In this way, he was able to centre the practitioners' perspective in his research and laid an important foundation for future research. Ed's account raised important questions about institutional and systemic limitations and opportunities to promote AR in doctoral programs.

Even as AR elements were brought into being in diverse ways, as AR PhD projects, all our research had to be materialized into single doctoral dissertations. Asha and Akina who are based in anthropology and sociology structured their theses based on the action-reflection cycles with their co-researchers. Abbie, Ming and Ed also presented the rich insights gained with their co-researchers in thesis format conventionally accepted in their respective disciplines of theatre studies and business, management and organizational research.

In conclusion, AR values and principles continued to shape our projects as our respective doctoral research journeys converged to form a distinct collaborative AR journey of its own. As we discussed and co-wrote in 'Collaborative 3.0' mode (Austin et al., 2021, p. 41) and engaged in 'triple loop learning' (Flood & Romm, 1995, 2018) of how, what and why of doing an AR PhD in light of our diverse and dynamic experiences, we began to see more fully our own projects, one another's projects and AR doctoral research more generally. We eventually found ourselves not merely researching and writing about our own projects but also taking concrete actions together to situate our AR PhD experiences within the wider field of AR. Two decades ago, Yoland Wadsworth published an article in ALAR journal in 2002 titled, *We are one (paradigmatic river) but we are many (tributary streams)*, which traced the evolving and expanding fields of AR (Wadsworth, 2002). She illustrated AR as an ever flowing, living and active paradigm made up of many different networks, streams and strands. In the Introduction of this Special Issue, Akihiro Ogawa traced the trajectory of a new AR stream he initiated when he came to Melbourne by connecting with AR scholars and networks and passing it on to his students. As AR was passed onto us, we realized we not only adopted AR into our doctoral projects, we also became an active part of the real-time making of this new stream. From this panoramic view of our research journeys interlacing as tributary streams with the paradigmatic river of AR, we conclude this Conclusion with the hopeful confidence that the collaborative reflections on our AR journeys presented here will inspire new students to go on their own AR journeys that will flow on to create new streams.

Acknowledgments

We would like to thank an anonymous reviewer for their helpful comments on an earlier version of this article. We especially thank them for pointing us to Gregory Bateson's (1987) work on third order of learning in our discussions on triple loop learning and Checkland and Holwell's (1998) idea of recoverability in our reflections on the seemingly endless nature of an AR project. Importantly, we would like to acknowledge the important contribution our colleague, Asha Ross, made to this AR PhD project. Although her final paper could not be submitted to the special issue following the first round of review, she played a vital role in our collaborative reflections on our AR PhD journeys and the co-writing process of this Conclusion. We miss you in this, Asha.

References

Austin, D, Bessemer, Y, Goff, S, Hill, G, Orr, L. & Vaartjes, V, 2021, 'Collaborative writing as action research: A story in the making', *Action Learning Action Research Journal*, vol. 27, no. 1, pp. 36-62.

Avison, DE., Lau, F, Myers, MD, & Nielsen, PA, 1999, 'Action research', *Communications of the ACM*, vol. 42, no. 1, pp. 94–97.

Bateson, G, 1987, *Steps to an ecology of mind: Collected essays in anthropology, psychiatry, evolution and epistemology*, Jason Aronson, Northvale, NJ; originally released 1972, reissued in 2000, University of Chicago Press, Chicago, IL. Available from: https://ejcj.orfaleacenter.ucsb.edu/wp-content/uploads/2017/06/1972.-Gregory-Bateson-Steps-to-an-Ecology-of-Mind.pdf [Accessed 29th August 2022]

Brydon-Miller, M, Greenwood, D, & Eikeland, O, 2006, 'Conclusion; Strategies for addressing ethical concerns in action research', *Action Research*, vol. 4, no. 1, pp. 129–131.

Checkland, P & Holwell, S, 1998, 'Action research: Its nature and validity', *Systemic Practice and Action Research*, vol. 11, pp. 9-21.

Davison, RM., Martinsons, MG, Wong, LHM, 2021, 'The ethics of action research participation', *Information System Journal*, vol. 32, no. 3, pp. 573-594. https://doi.org/10.1111/isj.12363

Erfan, A & Torbet, B, 2015, 'Collaborative developmental action inquiry', in Bradbury, H(Ed), *Sage handbook of action research* (3rd edn),Sage Publications, Thousand Oaks, pp. 64–75.

Flood, R.L & Romm, NRA, 1995, 'Diversity management: Triple loop learning', Systemic Practice and Action Research, vol. 8, no. 4, pp. 469-482. https://DOI:10.1007/BF02253396.

Flood, RL & Romm, NRA, 2018, 'A systemic approach to processes of power in learning organizations: Part 1 – literature, theory and methodology of triple loop learning, *The Learning Organization*, vol. 25, no. 4 pp. 260-272. https://doi.org/10.1108/TLO-10-2017-0101

Greenwood, D, 2002, 'Action research: Unfulfilled promises and unmet challenges', *Concepts and Transformation*, vol. 7, no. 2, pp. 117–139.

Haigh, C & Witham, G, 2015, *Distress protocol for qualitative data collection,*. Retrieved from https://www.mmu.ac.uk/media/mmuacuk/content/documents/rke/Advisory-Distress-Protocol.pdf. [Accessed 10 February 2017]

Jenkins, H, Ito, M & Boyd, D, 2016, *Participatory Culture in a Networked Era: A Conversation on Youth, Learning*, Polity Press, Cambridge.

Ngũgĩ wa Thiong'o, 1986, *Decolonising the mind: The politics of language in African literature*, James Currey, London.

Scott-Jones, J, 2010, 'Thinking through Ethnography: Introductions' in Scott-Jones, J & Watt, S (eds), Ethnography in Social Science Practice, Routledge, London, pp. 3-12. https://www.routledge.com/Ethnography-in-Social-Science-Practice/Scott-Jones-Watt/p/book/9780415543491 [Accessed 9 Feb 2022].

Schimmer, T, 2018, 'Growing 21st century learners: Assessing 7 critical Competencies', Webinar September 5, in *Solution Tree Webinar Series*. Retrieved from https://www.solutiontree.com/webinars/growing-21st-centurylearners- webinar.html. [Accessed 12 January 2022]

Tosey P, Visser, M & Saunders, MNK, 2012, 'The origins and conceptualizations of 'triple-loop' learning: A critical review', *Management Learning*, vol. 43, no. 3, pp. 291-307. https://doi.org/10.1177/1350507611426239.

Tuhiwai Smith, L, 2012, *Decolonizing methodologies: Research and Indigenous peoples*, Zed Books, London.

Wadsworth, Y, 2002, 'We are one (paradigmatic river) but we are many (tributary streams)', *Action Learning Action Research Journal*, vol. 7, no. 1, pp. 3–28.

Membership information and article submissions

Membership categories

Membership of Action Learning, Action Research Association Ltd (ALARA) takes two forms: individual and organisational.

ALARA individual membership

Members of the ALARA obtain access to all issues of the *Action Learning and Action Research Journal* (*ALARj*) twelve months before it becomes available to the public.

ALARA members receive regular emailed Action Learning and Action Research updates and access to web-based networks, discounts on conference/seminar registrations, and an on-line membership directory. The directory has details of members with information about interests as well as the ability to contact them.

ALARA organisational membership

ALARA is keen to make connections between people and activities in all strands, streams and variants associated with our paradigm. Areas include Action Learning, Action Research, process management, collaborative inquiry facilitation, systems thinking, Indigenous research and organisational learning and development. ALARA may appeal to people working at all levels in any kind of organisational, community, workplace or other practice setting.

ALARA invites organisational memberships with university schools, public sector units, corporate and Medium to Small Business, and community organisations. Such memberships include Affiliates. Details are on our membership link on our website (https://alarassociation.org/membership/Affiliates).

Become a member of ALARA

An individual Membership Application Form is on the last page of this Journal or individuals can join by clicking on the Membership Application button on ALARA's website. Organisations can apply by using the organisational membership application form on ALARA's website.

For more information on ALARA activities and to join

Please visit our web page:
https://www.alarassociation.org/user/register
or email admin@alarassociation.org

Journal submissions criteria and review process

The *ALARj* contains substantial articles, project reports, information about activities, creative works from the Action Learning and Action Research field, reflections on seminars and conferences, short articles related to the theory and practice of Action Learning and Action Research, and reviews of recent publications. *ALARj* also advertises practitioners' services for a fee.

The *ALARj* aims to be of the highest standard of writing from the field in order to extend the boundaries of theorisation of the practice, as well as the boundaries of its application.

ALARA aims *ALARj* to be accessible for readers and contributors while not compromising the need for sophistication that complex situations require. We encourage experienced practitioners and scholars to contribute, while being willing to publish new practitioners as a way of developing the field, and introduce novice practitioners presenting creative and insightful work

We will only receive articles that have been proof read, comply with the submission guidelines as identified on *ALARj*'s website, and that meet the criteria that the reviewers use. We are unlikely to publish an article that describes a project simply because its methodology is drawn from our field.

ALARA intends *AlARj* to provide high quality works for practitioners and funding bodies to use in the commissioning of works, and the progression of and inclusion of action research and action learning concepts and practices in policy and operations.

ALARj has a substantial international panel of experienced Action Learning and Action Research scholars and practitioners who offer double blind and transparent reviews at the request of the author.

Making your submission and developing your paper

Please send all contributions in Microsoft Word format to the Open Journal Systems (OJS) access portal: https://alarj.alarassociation.org.

You must register as an author to upload your document and work through the electronic pages of requirements to make your submission. ALARA's Managing Editor or Issue Editor will contact you and you can track progress of your paper on the OJS page.

If you have any difficulties or inquiries about submission or any other matters to do with ALARA publications contact the Managing Editor on editor@alarassociation.org.

For the full details of submitting to the *ALAR Journal*, please see the submission guidelines on ALARA's web site https://alarassociation.org/publications/submission-guidelines/alarj-submission-guidelines.

Guidelines

ALARj is devoted to the communication of the theory and practice of Action Learning, Action Research and related methodologies generally. As with all ALARA activities, all streams of work across all disciplines are welcome. These areas include Action Learning, Action Research, Participatory Action Research, systems thinking, inquiry process-facilitation, process management, and all the associated post-modern epistemologies and methods such as rural self-appraisal, auto-ethnography, appreciative inquiry, most significant change, open space technology, etc.

In reviewing submitted papers, our reviewers use the following criteria, which are important for authors to consider:

Criterion 1: How well are the paper and its focus both aimed at and/or grounded in the world of practice?

Criterion 2: How well are the paper and/or its subject explicitly and actively participative: research with, for and by people rather than on people?

Criterion 3: How well do the paper and/or its subject draw on a wide range of ways of knowing (including intuitive, experiential, presentational as well as conceptual) and link these appropriately to form theory of and in practices (praxis)?

Criterion 4: How well does the paper address questions that are of significance to the flourishing of human community and the more-than-human world as related to the foreseeable future?

Criterion 5: How well does the paper consider the ethics of research practice for this and multiple generations?

Criterion 6: How well does the paper and/or its subject aim to leave some lasting capacity amongst those involved, encompassing first, second and third person perspectives?

Criterion 7: How well do the paper and its subject offer critical insights into and critical reflections on the research and inquiry process?

Criteria 8: How well does the paper openly acknowledge there are culturally distinctive approaches to Action Research and Action Learning and seek to make explicit their own assumptions about non-Western/ Indigenous and Western approaches to Action Research and Action Learning

Criteria 9: How well does the paper engage the context of research with systemic thinking and practices

Criterion 10: How well do the paper and/or its subject progress AR and AL in the field (research, community, business, education or otherwise)?

Criterion 11: How well is the paper written?

Article preparation

ALARj submissions must be original and unpublished work suitable for an international audience and not under review by any other publisher or journal. No payment is associated with submissions. Copyright of published works remains with the author(s) shared with Action Learning, Action Research Association Ltd

While *ALARj* promotes established practice and related discourse *ALARj* also encourages unconventional approaches to reflecting on practice including poetry, artworks and other forms of creative expression that can in some instances progress the field more appropriately than academic forms of writing.

Submissions are uploaded to our Open Journal System (OJS) editing and publication site.

The reviewers use the OJS system to send authors feedback within a 2-3 month period. You will receive emails at each stage of the process with feedback, and if needed, instructions included in the email about how to make revisions and resubmit.

Access to the journal

The journal is published electronically on the OJS website.

EBSCO and InformIT also publish the journal commercially for worldwide access, and pdf or printed versions are available from various online booksellers or email admin@alarassociation.org.

For further information about the *ALAR Journal* and other ALARA publications, please see ALARA's web site http://www.alarassociation.org/publications.

Individual Membership Application Form

This form is for the use of individuals wishing to join ALARA.
Please complete all fields.

Name

| Title | Given Name | | Family Name |

Residential Address

| Street | Town / City | Postcode / Zip |
| Country | | |

Postal Address

| Street | Town / City | Postcode / Zip |
| State | Country | |

Telephone

| Country Code | Telephone Number |

Mobile Telephone

| Country Code | Mobile Number |

Email

Email Address

Experience (Please tick most relevant)
- ☐ No experience yet
- ☐ 1 – 5 years' experience
- ☐ More than 5 years' experience

Interests (Please tick all relevant)
- ☐ Education
- ☐ Health
- ☐ Community / Social Justice
- ☐ Indigenous Issues
- ☐ Gender Issues
- ☐ Organizational Development

Are you eligible for concessional membership?
If you are a full-time student, retired or an individual earning less than AUD 20,000 per year, about USD 13,750 (please check current conversion rates), you can apply for concessional membership.

Do you belong to an organization that is an Organizational Member of ALARA?
If you are a member of such an organization, you can apply for the Reduced Membership Fee. Please state the name of the Organizational Member of ALARA in the box below.

Payment
We offer a range of payment options. Details are provided on the Tax Invoice that we will send to you on receipt of your membership application.

[POST billpay] [BPAY] [MasterCard] [VISA]

If you want to join and pay online, please go to https://www.alarassociation.org and click on the Membership Application button (lower right). Alternatively, please complete and return this form to us.

By Post
ALARA Membership
PO Box 162 Greenslopes
Queensland 4120
AUSTRALIA

By FAX
+ 61 (7) 3342 1669

By Email
admin@alarassociation.org

Annual Membership Fees (Please select one)

Full Membership
- ☐ AUD 143.00
- ☐ AUD 99.00
- ☐ AUD 55.00

Concessional Membership
- Developed Country ☐ AUD 71.50
- Emerging Country ☐ AUD 49.50
- Developing Country ☐ AUD 27.50

Reduced Membership Fee, as I belong to an Organizational Member of ALARA
- Developed ☐ AUD 71.50
- Emerging ☐ AUD 49.50
- Developing ☐ AUD 27.50

Organization's name

Privacy Policy

By submitting this membership form, I acknowledge that I have read, understood and accept ALARA's Privacy Policy https://www.alarassociation.org/sites/default/files/docs/policies/ALARA_PrivacyPolicy11_1.pdf.

ALARA will acknowledge receipt of your application and send you an invoice or receipt of payment. You will receive an email confirming activation of your account, and details on how you can access website functions.

Printed by Libri Plureos GmbH in Hamburg, Germany